The Old House
HOLIDAY and PARTY
COOKBOOK

HOW TO USE STATISTICS

HOW TO USE STATISTICS

Joe D. Megeath

Metropolitan State College

With the assistance of
Allen T. Peck
Arapahoe Community College

CANFIELD PRESS ⌀ **SAN FRANCISCO**
A Department of Harper & Row, Publishers, Inc.
New York   Evanston   London

Sponsoring Editor: Jerry Papke
Production Editor: Pat Brewer
Copy Editor: Don Yoder
Designer: Penny L. Faron
Illustrator: Paulette Hanson
Line Artist: Carl Brown

Library of Congress Cataloging in Publication Data

Megeath, Joe D   1939–
  How to use statistics.

  Includes bibliographies.
  1. Statistics. I. Title.
HA29.M45   519.5   74–28237
ISBN 0–06–385445–7

HOW TO USE STATISTICS

Copyright © 1975 by Joe D. Megeath

75 76 77 10 9 8 7 6 5 4 3 2 1

CONTENTS

PREFACE

THIS TEXT HAS been written as a basic statistics primer for people who have no prior knowledge of the field. The user may be a formal student or may be someone on the job needing a reference book at work. The material has been generally directed toward business applications, but persons in other disciplines will find it relevant and useful. The contents are well suited to a two-quarter course, a one-semester course, or with proper selection of topics, a one-quarter course.

The material assumes no more than a working knowledge of basic algebra. There are no mathematical derivations. The emphasis throughout is on introducing topics on an intuitive level and then developing these topics into formal statistical concepts. Purely academic presentations have been studiously avoided whenever possible.

Most people I have encountered use statistics as a drunk uses a light pole—for support rather than illumination. This could be due in part to the human disposition, but it is due also to a lack in our training. In most institutions, the statistics course has one of the highest "fatality" rates of the courses offered. Students drop the course, fail the course, and have been known to change their major to avoid the course. And among those who have successfully "passed" their statistics requirements, there appears to be a woeful lack of understanding of what they studied. I remember in particular a colleague who said, "Oh, yeah, I took a statistics course in graduate school. It had something to do with a bell-shaped curve." Good Lord! A graduate course in statistics and he remembers "it had something to do with a bell-shaped curve"—what a waste!

Each chapter in this book begins with a set of general objectives that will alert the reader to what he or she should be looking for in the chapter. The problems are carefully coordinated with these objectives and cover a broad range of difficulty. There are problems for students who need a helping hand with basic concepts, and there are problems that are geared to push the reader to transfer his or her understanding of concepts to entirely new situations. At the end of each chapter is a summary of information and equations to aid the reader in reviewing the topics covered. Some chapters also contain an autopsy, which is designed to point out particular pitfalls or important subjective areas. The objectives, summaries, autopsies, and the motivating light-hearted examples and problems make this book very appropriate for self-paced courses.

The reader of this book will also be challenged to go beyond the cut-and-dried syndrome that can plague statistics. The fact of the matter is that the *use* of statistics is not cut and dried; there are many subjective decisions to be made by the analyst. I fully realize that it is far easier to ignore these subjective decisions for both the student and the instructor. However, practical experience has indicated that the easiest route is not necessarily the best route.

I have taught statistics to students of business, engineering, education, behavioral sciences, and mathematics. It has been my experience that *more* students get a *better* understanding of the subject if it is heavily seasoned with realistic examples and lively problems. Consequently, you will occasionally find that you don't know whether to laugh or to cry as you read. This is a noticeable improvement, though, over definitely knowing that you want to cry.

I would like to acknowledge the many students and friends who helped in a hundred different ways in the preparation of this text. Particularly, I would like to thank Paulette Hanson for her initial work without promise of pay; Howard Flomberg for the major share of the computer programming needed; and my wife, Donna, for putting up with a lot during this writing. This book might not have happened without the inspiration provided by my friends Irv Forkner, R. E. D. Woolsey, and John Rushton.

Joe D. Megeath
Denver, Colorado
January 1975

PART I
HOW TO DESCRIBE DATA

CHAPTER 1
STATISTICAL
AWARENESS

OBJECTIVES: After studying this chapter and working the problems, you should be able to:
Demonstrate with a current example your optimistic skepticism toward statistical data.
Explain why figures don't necessarily tell all the truth.

IN TODAY'S WORLD, including both our work world and our private world, we are besieged with numbers. We are exposed to long lists of sales figures, charts of costs, and such statements as:

"Our profits are 1 percent."

"A survey shows 55 percent of the voters support the bond issue."

"The average income of our salesmen is $20,000."

Statements like these can be heard or read almost every day. These three comments have two things in common: they all involve numbers and not one of them means anything standing by itself. Take a look at the statement, "We couldn't possibly be contributing to inflation. Our profits are only 1 percent." What does that mean? One percent of what? Is it a 1 percent return on investment? That's hardly likely in view of the fact that you can invest your money safely in a bank for far more than a 1 percent return. Perhaps the man meant 1 percent of sales. If that's the case, then we'd better look more closely. If I loan you 99 cents and at the end of the day you pay me back $1, then my profit of 1 cent is 1 percent of the amount you paid me. If we do this every day, however, I'll make a penny each day and at the end of one year I will have made $3.65. Putting that in terms of a return on my investment of 99 cents, I will have made $3.65/.99 = 369$ percent return on my investment. The 369 percent profit (percentage of return on investment) looks a lot different than the 1 percent profit (percentage of sales), doesn't it? The point is that by looking at "1 percent profit" the amount of information you received was totally nothing. You still have no idea whether or not the profit was relatively high or relatively low.

Suppose in your community there is a controversial bond issue that would be used to build a new ten-lane, fiber glass bridge to replace the current footbridge over Little Dry Gulch at the edge of town. One morning you open the newspaper and the headline screams, "Survey Shows 55 percent of Voters Support Bond Issue." You might think, "Well, people seem to support this issue. I guess it must be a good deal." You might think, though, "Says who?" Perhaps the last sentence of the lengthy article states that the survey of 100 voters was conducted by the National Bridge Builders Association. That should lead you to conclude that the survey results, by themselves, tell you nothing. The Bridge Builders Association has a heavy interest in the project, and a survey can be gerrymandered in a lot of ways—either consciously or subconsciously. For instance, the people surveyed could have included all major suppliers of bridge-building materials in the area. Or the question could have been leading, such as "Do you support the new bridge bond or do you want to see your child drown in Little Dry Gulch?" These examples may seem extreme, but you should be wary of surveys if they are not conducted by a neutral party.

In the statement "The average income of our salesmen is $20,000," the implication to the lay reader is that most of the salesmen make about $20,000. If one salesman makes $100,000 and four salesmen make $0, however, the average income for the five is $20,000. Such general statements don't tell you anything by themselves because the word "average" can imply something different than it actually means.

If you get nothing else out of a study of statistics, you should learn to be optimistically skeptical of quoted numbers. This attitude of optimistic skepticism and an increased awareness of numbers is what I call *statistical awareness.*

Once you develop statistical awareness you will find that most numbers you hear or read tell you very little without further investigation. Numbers are like the Bible in one sense. When they are quoted, they add credence to whatever is being said. But also, like the Bible, they are often quoted erroneously or out of context.

People often talk about football statistics or about something becoming a statistic, referring to actual counts and figures that are compiled for various reasons. This usage of the term has led many people to assume that the field of statistics deals with counting the number of births, recording the number of traffic accidents, adding the total yardage gained, and so forth. Such numbers are *statistical data*.

The term *statistics* refers here to a much larger scope, including methods of analyzing such data. These methods encompass techniques for summarizing and describing statistical data (descriptive statistics) and techniques for making inferences from sample data (inferential statistics).

The businessman today, large or small, receives statistical data in huge amounts. If he cannot get the correct information from those data, his business will suffer. To be statistically illiterate is to be at the mercy of any number (or lack of numbers) you happen to come across.

To protect yourself against "tell you nothing" data and to reach the proper decisions and actions from data, you must be able to handle data almost like a witness in a courtroom. A witness is sworn to tell the truth, the whole truth, and nothing but the truth. You must develop the ability to place statistical data on the witness stand and find out what information they really tell you, all the information they contain, and nothing more than the information they contain.

This ability requires you to ask pertinent questions of the number in the witness chair. If you read that "the unemployment rate is 5 percent," for instance, you should ask:

Percentage of what?

Does this include people who are seeking temporary employment?

Does this include people who have obtained employment but haven't begun work yet?

What was this percentage a year ago?

Is this a national or a local figure?

Are there any extenuating circumstances that could cause this figure to be so high (or low)?

Who compiled the data?

The first step in putting statistical data on the witness stand is to develop a statistical awareness. Ask yourself what the numbers really mean and "says who?" The second step is to develop some technical skill and learn definite procedures for cross-examining the data in the witness chair. These procedures to be learned constitute the field of statistics.

The major point, though, is making numbers "talk" to you. Many people think that a number is a number is a number is a number. But that isn't necessarily true. Take the number 2. If I say, "I have two cents," the listener will yawn and think so what. But if I say, "I have two wives," the listener is immediately going to raise an eyebrow . . . at least! It is the same number in both statements, yet it obviously carries a different impact. (Haven't you ever been in a conversation where someone says, "Do you realize we sold ten purple thingamajigs yesterday?" And you don't know whether to say "That's too bad" or to say "That's great"?)

At other times we become immune to numbers. For instance, the cost to the United States of the war in Indochina was reported as $108 billion.* How big is 108 billion? That is, written out, 108,000,000,000. We often accept such figures because we don't grasp their magnitude. If a baby started counting one digit each second at the moment of his birth and passed this task on to his children, it would take more than 48 generations to count to 108 billion!

As you develop your ability with formal statistical techniques, you will also be (probably subconsciously) developing your statistical awareness. Start right now by looking more closely at the numbers being used around you. Some people claim that figures don't lie, but liars figure. You will begin to find that figures don't necessarily tell all the truth either.

PROBLEMS

1.   Watch for a quoted or written figure that is misleading or unsubstantiated. Analyze why the figure tells you nothing.

* "The Fighting Finally Stops for the U.S.," *Time*, 27 August 1973, p. 34.

2.   Watch for a quoted or written figure that you think is not mis-leading and is totally substantiated. Discuss it with friends to see if they agree with you.

3.   Assume you are the police commissioner in your community. A newspaper headline proclaims that there were 1250 thefts during the last six months.

> *a.*   Not a pleasant figure, but what does it tell you?
>
> *b.*   What if there were 1100 thefts during the previous six months?
>
> *c.*   Would your reaction be any different if your town's population was 10,000 or 500,000 people? Why?
>
> *d.*   How would crime rate figures from a similar city help you decide whether or not to fire the police chief?
>
> *e.*   Suppose your city could get a federal grant of $10,000 to put a special antitheft squad on the street. To get the grant your city would have to put up $5000 and show a theft reduction of 20 percent. What questions would you ask before you made a decision?

SELECTED REFERENCES

Campbell, Stephen K. *Flaws and Fallacies in Statistical Thinking.* Englewood Cliffs, N.J.: Prentice-Hall, 1974.

Huff, Darrell. *How to Lie with Statistics.* New York: Norton, 1954.

CHAPTER 2
MAKING
DATA USABLE

OBJECTIVES: After studying this chapter and working the problems, you should be able to:

Explain why the proper presentation of data is important.

Explain how and why ratios are used.

Discuss when to use a table and when to use a chart to display data.

Prepare examples of charts and tables (bar chart, line chart, cross-classification table, and others).

Illustrate why scale selection and area proportioning for charts are important.

DATA DON'T ALWAYS come as a single number or a pair of numbers. Often they appear as a crowd. When this happens, we need to organize them into a usable format. By usable format, I mean that the full impact of the numbers becomes apparent. This normally requires a table, graph, or chart. A look at the construction of tables, graphs, and charts will benefit the would-be analyst in two ways: he will learn how to prepare them and how to read them.

Preparing good tables, graphs, and charts requires a mixture of general technical knowledge about what you should and shouldn't do with them, common sense, and practice. There are two cardinal rules in preparing them. Observance of these two rules will help keep you out of trouble when you are displaying data:

Rule 1: Answer the question: What is the precise purpose of this table or graph or chart?

Rule 2: Make the table or graph or chart stand by itself so that your readers do not have to refer to something else to find out what it means.

RATIOS VERSUS QUANTITIES

Quantities give us such information as the production last year or the dollar sales volume for each month. When we want to make comparisons or analyze data, though, we often must convert them into some kind of pertinent ratio. If we want to compare the performance of plant A and plant B, for instance, the *actual* production for the month may merely reflect that plant A either worked more days or is a bigger plant. Line 1 in Table 2-1 shows the actual production.

TABLE 2-1.   Plant A Versus Plant B

	A	B
1. Actual production (units)	100,000	75,000
2. Units of production per working day	3333.3	3000.0
3. Units of production per man-hour of labor	301.0	322.2

To give a better measure of the relative performance, the ratio of units of production to number of working days is shown in line 2. This ratio still shows plant A to be better, but not so much as in actual production. An even more accurate measure of their relative performance may be found by converting the actual units of production to the ratio of production per man-hour of labor as shown in line 3. This indicates that plant B may actually be the more efficient of the two—if both have the same degree of automation, of course. The choice of a proper ratio can bring to light many facts that the actual or raw numbers disguise.

TABLES

We will use the term *tables* to mean an array or matrix of numbers
that are split up or identified by certain characteristics or categories
(sales by year, employees by sex, employees by income level, sales by
product line, and so forth). Graphs and charts, which we will discuss
later, are a pictorial representation of data rather than the absolute
figures that appear in a table.

A table is used when exact figures are necessary. Sometimes the
purpose of a table is to give general information, as do the tables of
data found in the Census Surveys. When a table is being used to make
comparisons or to make an analysis instead of to present general
information, however, the analyst should try to keep the number of
figures to a minimum. There is generally a strong impulse to add
"one more column of figures just in case anyone is interested." The
result of adding those extra columns is generally a table which so
overwhelms the reader that the impact is lost. If the table must contain a lot of figures, the analyst should highlight the important elements with circles or arrows or appropriate symbols to guide the
reader's attention.

Although many people feel that the construction of tables is almost
child's play, there is considerable art to it. It is quite easy to bury
significant facts in a table. Consider a situation in which a company
has decided to market decorative garden lights. They have three
designs under consideration: one that looks like dandelions, one that
looks like sick sunflowers, and one that looks like droopy daffodils.
Of course, the original artist didn't design them like that, but as often
happens, that's the way the prototypes came out of the production
process. They want to produce only one design, though, because this
would greatly reduce production and inventory cost. A market survey

TABLE 2-2.   "Which of these three designs do you
like most?"

Design	Responses
Dandelions	40
Sunflowers	50
Daffodils	10
None	0
	100

TABLE 2-3. "Are there any of these designs that you think are terrible?"

Design	Responses
Dandelions	39
Sunflowers	41
Daffodils	20
None	0
	100

team takes mockups of the three designs to a sample of consumers. Included in the questions is "Which of these three designs do you like most?" Table 2-2 displays their responses.

Table 2-2 seems to indicate that the company should produce the sunflower design. However, a back-up question (which is usually advisable) was included in the questionnaire: "Are there any of these designs that you think are terrible?" The responses are shown in Table 2-3. This table creates a conflict. The daffodil design shows the best response on the "terrible" question and our choice thus far, the sunflower design, has a 41 percent response, which is relatively high.

A third table is then made from the questionnaire to "cross-classify" the two questions. A cross-classification categorizes data by two characteristics simultaneously. In this cross-classification table (Table 2-4) we see, for instance, that of the 39 respondents who thought the dandelion design was terrible, 31 were people who liked the sunflower design most. The dashes on the diagonal indicate that, obviously, no one could like a design most and also least.

TABLE 2-4. Number of Responses to Questions of "Like Most" versus "Terrible"

| Terrible | Like Most | | | Total |
	Dandelions	Sunflowers	Daffodils	
Dandelions	—	31	8	39
Sunflowers	39	—	2	41
Daffodils	1	19	—	20
	40	50	10	100

This table brings to light the fact that the survey indicates a distinct market segmentation. Those consumers who liked the dandelions most would not buy the sunflower design, while those who liked the sunflowers most would not buy the dandelion design. This table tells the analyst that the company should reconsider their original position of marketing only one design. The responses in Table 2-4 indicate that there would be very little "switch" buying. In other words, if the sunflower design were the only one to be produced and marketed, the company should not anticipate that many people who like dandelion designs would purchase the sunflower design. On the other hand, if both sunflower and dandelion designs were produced and marketed, they would anticipate a greater total volume of sales.

GRAPHS AND CHARTS

Graphs and charts should be used to display data when a quick visual picture is appropriate, when the data need not be conveyed with pinpoint accuracy, or when trends are to be emphasized. Charts can have a much higher impact than a table on a reader because of their visual effect. However, they must be constructed carefully because the visual effect can be easily misread or misused.

A *line graph* is made by connecting the data points with a line and is most often used to illustrate changes or trends over time. A *bar chart* uses vertical or horizontal bars or rectangles to illustrate magnitude. It can also be used to illustrate changes or trends over time, but is more commonly used to illustrate comparisons between items. A *component chart* is a special bar chart: each bar is partitioned into components to illustrate how much of the total is allocated to various parts. Another special bar chart is a *histogram*—a set of vertical bars each of which represents the frequency of occurrence of a certain value. A histogram is an excellent way to portray the relative frequency of observed values. A *pictogram* is a special chart that uses symbols or facsimiles drawn to scale to illustrate magnitudes.

> **Example 1:** To demonstrate some tables, graphs, and charts, let's look at Crab's Tire Company, which operates three retail stores. They are interested at this time in the sales of premium tires. Table 2-5 is a general information table that has been drawn from past sales records.

TABLE 2-5. Premium Tire Sales by
Year by Store

Store	1972	Year 1973	1974
A	900	1200	1500
B	600	1050	1400
C	500	550	1100
	2000	2800	4000

From Table 2-5, a number of displays could be drawn. If we wanted to illustrate the growth of the stores over the past three years, we could prepare a line graph such as the one shown in Figure 2–1.

FIGURE 2–1. Premium Tire Sales by
Year by Store

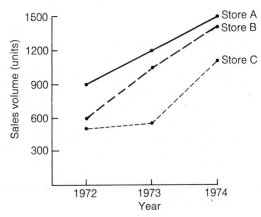

Remember that in line charts the scale must be relative and constant. For instance, if the manager of store A wanted to leave the visual impression of an "explosive" sales record, he might want a chart such as the one shown in Figure 2–2. This chart presents a picture of more sharply rising sales than the actual facts and data can support. Notice that two things have been done in this chart: (1) the vertical scale is broken between 0 and 900,

FIGURE 2-2.
Premium Tire Sales for
Store A: 1972-1974

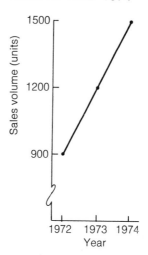

indicating missing points; and (2) the horizontal scale
has been shortened to an extreme. Both these items make
the line indicating sales rise at a sharp angle — and both
practices should be avoided in making charts. A chart
contains both visual impact and actual data (as shown on
the scale). However, the visual impact will generally over-
ride the data shown. It is the visual impression that sticks
with the reader, not the actual data. In Figure 2-2, the

FIGURE 2-3.   Premium Tire Sales for Store A:
1972-1974

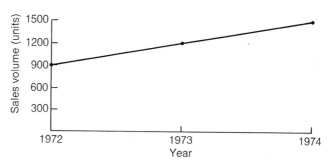

FIGURE 2–4. Premium Tire Sales by Store by Year

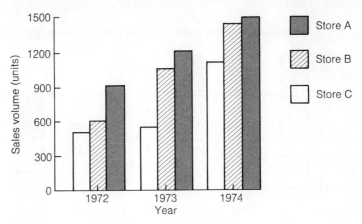

initial visual impression is that sales have doubled or tripled over the past three years.

Scales can be altered to create the opposite illusion also. Perhaps the manager of store A wants to leave the impression of a slow, steady growth so that nobody will ask him why profits haven't gone up more than they have. He would then want to get away with presenting the sales data as in Figure 2–3. The visual impression here is of slow, steady growth. Notice, though, that the horizontal axis has been elongated. If the horizontal scale is lengthened relative to the vertical scale, the changes from one point to the next will appear to be slight.

If we were interested in comparing stores from time period to time period, we might want to construct a bar chart rather than a line chart. Such a bar chart is shown in Figure 2–4. In constructing bar charts, you should be careful to make the *areas* of the bars rather than the heights of the bars proportional to each other. If the widths of the bars are all the same, this will have no effect. But if one bar is twice as wide as another for some reason, the height should be reduced by one-half so that the areas are proportional. This concept will be discussed in more detail when we look at pictograms.

We may be interested in conveying how the stores have contributed to total sales in premium tires. If this is the case, we would use a component chart. Figure 2–5 shows

FIGURE 2–5.   Premium Tire Sales by Store by Year:
1972–1974

Store sales as % of total sales   Store contribution to total sales

Store A   Store B   Store C

two different ways we can construct a component chart:
as a percentage of total or as a part of total sales volume.

If we were interested in graphically illustrating the
company's increase in premium tire sales, a pictogram
might be the most appropriate means. A bar chart can be
used for the same purpose, but a pictogram has much
more eye appeal. Figure 2–6 would be an appropriate
pictogram: here a sketch of a tire illustrates volume.
In Figure 2–6 the *areas* of the two symbols are in the correct
proportions. The tire for 1972 has exactly one-half the

FIGURE 2–6.   Premium Tire Sales: 1972 Versus 1974

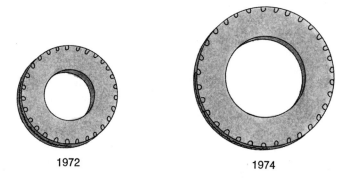

1972   1974

FIGURE 2–7.   Premium Tire Sales: 1972 Versus 1974

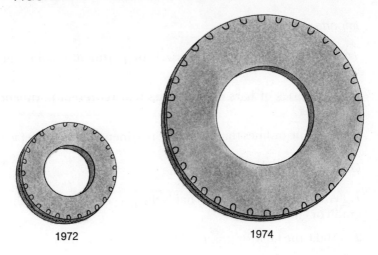

1972   1974

area of the tire for 1974, and this is the proportion of the
two sales figures: 2000 versus 4000.

Sometimes pictograms are erroneously used as in
Figure 2–7. Here the radius for 1972 is exactly one-half
the radius for 1974. The area of a circle, though, is $A =
\pi r^2$. Therefore the tire for 1974 has four times the area
for 1972. Whether you know the formula for the area of a
circle or not, your eye will pick up the information that
1974 is four times larger than 1972!

AUTOPSY

A lot of the data you present (or have presented to you) are in the
form of a table, graph, or chart. These visual displays represent
an excellent way of organizing data and presenting them in a readable
fashion so that the message is obvious. As you become more optimistically skeptical about the charts you read, you will find that a
careful reading of the data does not necessarily support your initial
impression from the chart. You will also occasionally find a chart
that is pretty to look at but totally unreadable when you try to find
specific facts from it (this is usually in the form of an exotic three-
dimensional figure in full color).

SUMMARY

Important Terms

Table: an array of data categorized by pertinent qualities or characteristics

Chart: a series of bars or rectangles that represent pertinent quantitative facts

Graph: a line or lines that represent pertinent quantitative facts

Overall Rules for Displaying Data

1. Answer the question: What is the precise purpose of this table or graph or chart?

2. Make the table or graph or chart stand on its own.

Specific Rules for Displaying Data

1. Don't crowd too much into an individual display.

2. Keep the scales in a chart constant and relative.

3. Make figures or symbols proportional in *area* to give the correct impression of the data proportions.

PROBLEMS

Note: These problems are designed to teach you the sources of data as well as how to display them. Some problems will require that you search out certain data.

1. Find an example of a misleading chart from a magazine or newspaper. Discuss why it is misleading and how it could be corrected.

2. Suppose that over the four years of 1971 through 1974 the dollar sales of Ace Company were $975,075, $1,030,140, $1,159,214, and $1,344,447, respectively. Over the same period the expenditures were $1,000,475, $1,001,222, $945,738, and $1,127,117, respectively. The capital invested in the business for each of these four years was $900,000, $950,000, $975,000, and $925,000.

> *a.* Make a display of this information for a financial analyst.

   b. Make a display of this information to show during a
   luncheon talk at a local civic organization.

   c. Why would a table be better for part (*a*) and a chart for
   part (*b*)?

3. Draw a line graph for the dollar sales figures in Problem 2
(*a*) using a scale of 1 inch = $1000 on the vertical axis and 1 year =
¼ inch on the horizontal axis; (*b*) using a scale of 1 inch = $2 million
on the vertical axis and 2 inches = 1 year on the horizontal axis; and
(*c*) using a scale of 1 inch = $500,000 on the vertical axis and 1 inch =
1 year on the horizontal axis. Which of these three scales do you
think would be the most appropriate to use?

4. From census data found in most libraries, determine the population of your state and that of the two states closest to yours in
population in 1940. Draw a chart that shows the comparative population changes in these three states from 1940 through 1970.

5. Construct a table and a chart that compare the gross national
income with your annual income over the last five years.

6. Contact a company in your area and find their total sales annually
for the last four years. Depict the first year's sales and the fourth
year's sales in a pictogram.

7. Construct a component chart for the types of expenditures
made by the federal government in the last full year and the year
previous.

8. Construct a table for new car sales in your state by major manufacturer.

9. The following table shows the premium receipts of United States
life insurance companies (in millions of dollars):

	1966	1967	1968	1969	1970
Life insurance premiums	17,160	18,094	19,364	20,491	21,679
Annuity premiums	2,416	2,671	2,993	3,762	3,721
Health premiums	7,244	7,887	8,730	9,743	11,367

Source: Institute of Life Insurance.

   a. Construct a line chart for the three types of premium
   receipts.

b. Construct a bar chart comparing the three types of premium receipts for each of the years shown.

c. Construct a component chart reflecting the percentage of total receipts contributed by each type of premium for each year.

d. Construct a component chart reflecting the actual dollars contributed by each type of premium for each year.

e. Construct a pictogram to compare the *total* premium receipts of 1966 versus 1970.

SELECTED REFERENCES

Campbell, Stephen K. *Flaws and Fallacies in Statistical Thinking.* Englewood Cliffs, N.J.: Prentice-Hall, 1974. Especially Chapter 5.

Huff, Darrell. *How to Lie with Statistics.* New York: Norton, 1954. Especially Chapters 5, 6.

Tanur, Judith M., Frederick Mosteller, William K. Kruskal, Richard F. Link, Richard S. Pieters, Gerald R. Rising. *Statistics: A Guide to the Unknown.* San Francisco: Holden-Day, 1972.

CHAPTER 3
DESCRIBING CENTRAL TENDENCY

OBJECTIVES: After studying this chapter and working the problems, you should be able to:

Illustrate the terms population, sample, parameter, *and* statistic.

Calculate the median, mode, and mean for grouped and ungrouped data and interpret the meaning of each.

WE HAVE SEEN how we can arrange sets of data so that we can more easily and accurately get an overview of comparisons and pertinent facts. That aspect of describing data is very useful to us. Often, though, we do not want to look at the entire set of data—either because there are far too many numbers for us to absorb at one time or because we are more interested in certain characteristics of the numbers than in the numbers themselves.

Suppose, for instance, that we wanted to compare daily sales this year with daily sales last year. We could make a chart with 365 entries such as the following:

Day	1972	1973
Jan. 1	$100	$ 98
Jan. 2	54	75
Jan. 3	103	102
.	.	.
.	.	.
.	.	.
Dec. 31	231	254

It doesn't take much imagination to realize that looking down that list of 365 lines is going to be laborious. And when we get through looking down the column of 365 lines, we are still going to be uncertain about what it told us. What we need to develop, then, is a systematic procedure for describing data without *listing* all the numbers in the data set.

MEAN, MEDIAN, AND MODE

There are some valuable calculations we can make to summarize a set of data rather than listing them. This chapter concerns itself with describing the midpoint or average or central tendency of the data. Specifically, we will look at the mean, the median, and the mode.

THE MEAN: THE BALANCE POINT

You probably have at least an intuitive idea of the mean already. The mean of a group of data is found by adding the data and then dividing the sum by the number of items in the group.

The mean or arithmetic average of a group of data can be thought of as the balance point: if you arranged the data on a horizontal scale, the mean would be the point at which the numbers on the left "balance" the numbers on the right. The average of 1, 4, 5, 0, 5 is found in two steps:

1. Adding the data:

$$
\begin{array}{r}
1 \\
+4 \\
+5 \\
+0 \\
\underline{+5} \\
15
\end{array}
$$

2. Dividing the sum by the number of items in the group (there are five items here):

$$
\begin{array}{r}
3 \\
5\overline{)15}
\end{array}
$$

To see that the mean is the balance point, suppose these numbers were arranged on a board with a consistent scale as shown in Figure 3–1. We have equal weights at the points 0, 1, and 4, and two such weights at 5. The fulcrum of weights arranged accordingly would be

FIGURE 3–1. Arithmetic Average of the Numbers 1, 4, 5, 0, 5

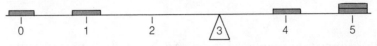

at the mean, 3, as shown in the figure. The mean can be thought of as the balance point in this sense.

In statistical terminology we say that the mean, denoted by $\bar{x}$ (x bar), is

$$\bar{x} = \frac{\sum_{i=1}^{n} x_i}{n}$$

That funny little squiggle, Σ, is the Greek symbol we use to mean "sum of." The Σ is part of summation notation. The i's refer to a subscript (a particular observation) and n is the total number of items we will look at. In other words, in summation notation our example would have appeared as

$$x_1 = 1 \qquad \text{first observation}$$

$$x_2 = 4 \qquad \text{second observation}$$

$$x_3 = 5 \qquad \text{third observation}$$

$$x_4 = 0 \qquad \text{fourth observation}$$

$$x_5 = 5 \qquad \text{fifth observation}$$

and

$$\frac{\sum\limits_{i=1}^{n} x_i}{n} = \frac{x_1 + x_2 + x_3 + x_4 + x_5}{5}$$

$$= \frac{1 + 4 + 5 + 0 + 5}{5}$$

$$= \frac{15}{5}$$

$$= 3$$

In summation notation the $i = 1$ in $\sum\limits_{i=1}^{n}$ tells us to start adding at the first observation, x_1, and the n above the Σ tells us to keep adding through n observations or up to x_n. In basic statistical work we are usually adding all the observations, so we commonly abridge $\sum\limits_{i=1}^{n} x_i$ to merely Σx, deleting the $i = 1$, the n, and the i as a subscript on x. This makes it much easier to write, and whenever you see Σx, remember that it is just shorthand for $\sum\limits_{i=1}^{n} x_i$.

We need to make one more distinction here. When we are talking about the mean of a *population*, we use the Greek symbol μ (mu): $\mu = \Sigma x/N$. A population is the total group or set with which we are concerned. The number of items in the total group or population is denoted by capital N. The mean μ of a population is a *parameter* of a population. A parameter is a characteristic of a population. When we are talking about the mean of a *sample*, we use the symbol $\bar{x}$: $\bar{x} = \Sigma x/n$. A sample is a part, a subset, or a subgroup of a population. The number of items in a sample is denoted by small n. The mean $\bar{x}$ of a sample is a *statistic* of a sample. A statistic is a characteristic of a sample.

If we were concerned about the average number of words per page in this book, then the population would be the number of words on each and every page of this book. If we were willing to estimate this average by examining ten pages, however, the number of words on each of these ten pages would represent a sample or a subset of the population. A summary of this notation is shown in Table 3-1. We will add to this list in later chapters. The notation used here is fairly standard in statistical works.

During the discussion of statistical methods it is often necessary to go through a number of formulas to clarify the meaning of a par-

TABLE 3-1. Statistical Notation

Item	For a Population	For a Sample
Number of items	N	n
Mean	μ	$\bar{x}$

ticular concept. Students are occasionally confused, then, as to which formula is the final one or the one to be used. I will identify the final formula, or the one with which you should be concerned, by the notation "Should Observe and Believe." To save space we will just use the initials of this notation. Therefore, thus far we have two final or usable equations:

SO&B 3-1:
$$\mu = \frac{\Sigma x}{N}$$

and

SO&B 3-2:
$$\bar{x} = \frac{\Sigma x}{n}$$

You may be interested in measures of central tendency if you are considering purchasing the Veni, Vidi, Vendo Company (their motto is "I came, I saw, I sold"), which sells its products through coin-operated vending machines. The vending machines dispense gum for 5 cents (a nickel), coffee for 10 cents (a dime), or an antacid tablet for 25 cents (a quarter). The machines do not give any change nor will they accept anything but one coin per purchase. In other words, two nickels cannot be used to purchase coffee.

You, as a potential purchaser of this apparatus, are interested in knowing the average amount spent by a customer. A customer here means a person who buys one item. If he purchases another item, he is counted as another customer. The population here is the purchases of *all* the customers of the company's machines. Obviously, this is too large a group with which to work, so you will only look at a sample of customer purchases. You are, in particular, interested in the *average* amount of a purchase made by a customer. Disregarding the smallness of the sample, suppose you watch one vending machine for an hour and observe ten customers during that period. The purchases they make are recorded as:

Customer (i)	Purchase (x_i)
1	5¢
2	10
3	5
4	25
5	5
6	5
7	−45
8	5
9	10
10	5
	30¢

The −45 cents happened when a little old lady in hobnailed boots bought gum for a nickel and then kicked the machine in the lower right-hand corner and got two quarters back.

To find the mean purchase amount we calculate

$$\bar{x} = \frac{\Sigma x}{n}$$

$$= \frac{30}{10}$$

$$= 3¢$$

The mean, $\bar{x} = 3$ cents, is our first rough estimate of the amount purchased per customer. If you feel that this sample of ten is representative of all customers, then you can merely count the number of customers and multiply by 3 cents to estimate the total amount of revenues for that group of customers. We will discuss the mean again in this chapter when we compare it to other measures of central tendency. Right now, let's look at the median.

THE MEDIAN: THE HALFWAY POINT

An alternative method of describing the central tendency of data is the median. The median is very simple in nature—it is merely the

halfway point of the data. In other words, one-half of the observations
will be less than the median and one-half will be greater.

To find the median, first put the data in ascending or descending
order. For the numbers 1, 4, 5, 0, 5 we would arrange them like this:

0

1

4 ← Median

5

5

The number 4 is such that there are two observations above (smaller
than) it and two observations below (larger than) it. The halfway
point of 4, then, is the median of this group of data.

If you prefer a formula to rely upon instead of just figuring out
where the halfway point is, you can calculate for an *odd* number
of observations:

SO&B 3-3:   Median $=$ value of the $\left(\dfrac{n+1}{2}\right)$th observation

In this example there is an odd number of observations ($n=5$), so the

$$\left(\frac{n+1}{2}\right)\text{th observation} = \left(\frac{5+1}{2}\right)\text{th observation}$$

$$= 3\text{rd observation}$$

and the value of the third observation (*after* they have been arranged
in ascending order) is 4.

If we are presented with an even number of observations, we treat
it slightly differently. For instance, in the group 1, 4, 5, 0 we first
arrange them in ascending order as before:

0

1
← Median
4

5

Notice that the median or halfway point is now between the second observation, 1, and the third observation, 4. We take the median to be that value halfway between these two observations: median = $(1 + 4)/2 = 2.5$. To calculate the median of an even number of observations, we find

SO&B 3-4: $$\text{Median} = \frac{\text{value of } (n/2)\text{th observation} + \text{value of } (n/2 + 1)\text{th observation}}{2}$$

An important characteristic of the median is that, unlike the mean, it is not affected by extreme values. Looking again at our example of the Veni, Vidi, Vendo Company, we observed purchases of 5, 10, 5, 25, 5, 5, −45, 5, 10, 5. The extreme value of −45 cents had a considerable effect on the mean. To find the median of this set of data, we first arrange the numbers in ascending order:

Observation	Amount
1st	−45¢
2nd	5
3rd	5
4th	5
→ 5th	5
→ 6th	5
7th	5
8th	10
9th	10
10th	25

$$\frac{5 + 5}{2} = \frac{10}{2} = 5$$

There is an even number of observations ($n = 10$), so (using SO&B 3-4)

$$\text{Median} = \frac{\text{value of } (n/2)\text{th observation} + \text{value of } (n/2 + 1)\text{th observation}}{2}$$

$$= \frac{\text{value of 5th observation} + \text{value of 6th observation}}{2}$$

$$= \frac{5 + 5}{2}$$

$$= \frac{10}{2}$$

$$= 5¢$$

Note that the median of 5 cents is higher than the mean $\bar{x}$ of 3 cents. This is due primarily to the fact that the full impact of −45 cents is felt in the calculation of the mean, but in the median the −45 cents is just another observation. The median becomes a valuable descriptive statistic to us in cases where a few extreme values would give us a distorted idea about the population. Before we make these comparisons, though, let's look at a third measure of central tendency.

THE MODE: THE MOST COMMON POINT

The mode is the simplest of the measures of central tendency. The mode is merely the value that occurs most often. In our observations on the Veni, Vidi, Vendo Company of 5, 10, 5, 25, 5, 5, −45, 5, 10, 5, we could count the number of occurrences of each value as follows:

Value	Frequency	
−45¢	I	(1)
5	JHT I	(6)
10	II	(2)
25	I	(1)

The value of 5 occurs six times, so it is the mode. The mode can be a useful description of the central tendency when one value occurs in a disproportionately high frequency. The mode can also bring to light another situation referred to as *bimodal*. Bimodal means that a set of data has two values which occur most often.

Suppose, for instance, that we are looking at the number of hours of overtime which employees worked in one particular week. Our frequency chart may look like Table 3-2.

Here we see that the data have two modes—one at three hours of overtime and the other at seven hours of overtime. A bimodal set of data indicates that we may be looking at two different populations.

TABLE 3-2.   Overtime Worked
Per Employee: Week of June 11

Hours of Overtime	No. of Employees
1	0
2	2
3	6
4	3
5	2
6	4
7	6
8	1

The employees could be from two different departments having different overtime policies. Or there may be some who don't want to work overtime but are required to put in a few hours whereas another distinct group actively solicits overtime work. When a bimodal situation is observed, the analyst should dig into the facts behind the data to see whether two populations can be identified.

GROUPED DATA

We now need to expand our ideas of the mean and the median to include *grouped data.* By grouped data we merely mean data that have been summarized by value. Sometimes the amount of data is so great that we must summarize them either by counting frequencies of occurrence of the same value or by merely classifying observations as falling within a certain interval rather than the actual value of that observation.

THE MEAN—GROUPED DATA

We estimate the mean of grouped data by a measurement that is sometimes called the *weighted mean.* The weighted mean is merely a shorthand method of finding the mean when many of the observations have the same value. If we have two observations with the value of 6, then instead of adding $6 + 6$ we can multiply 2 times 6. If five observations have the value of 6, then instead of adding $6 + 6 + 6 + 6 + 6$ we can multiply 5 times 6. If we let x be the assigned value of the observations and let f denote the number of times that value was observed, then the mean of grouped data can be found by the formula

SO&B 3-5:
$$\bar{x} = \frac{\Sigma fx}{n}$$

where n is the total number of observations in the sample.

In the observations on the vending machine we could have grouped the observations instead of listing them individually. In that example the grouped data would appear as

Value (x)	Frequency (f)
−45¢	1
5	6
10	2
25	1
	$n = 10$

and the mean could have been calculated by

$$\bar{x} = \frac{\Sigma fx}{n} = \frac{1(-45) + 6(5) + 2(10) + 1(25)}{10}$$

$$= \frac{30}{10}$$

$$= 3¢$$

This, of course, is the same answer we obtained previously, but the calculation has been shortened considerably.

If we had gathered data concerning the amount of money collected in a machine each day for a sample of 25 days, we might have come up with observations such as those shown in Table 3-3.

TABLE 3-3.   Money Collected from Machine 204 Per Day

Daily Amount	Midpoint (x)	Frequency (f)
$35–$44.99	40	2
$45–$54.99	50	5
$55–$64.99	60	7
$65–$74.99	70	8
$75–$84.99	80	3
		$n = 25$

Using SO&B 3-5 we calculate the mean of these data as

$$\bar{x} = \frac{\Sigma fx}{n} = \frac{2(40) + 5(50) + 7(60) + 8(70) + 3(80)}{25}$$

$$= \frac{1550}{25}$$

$$= \$62$$

Note that when we use grouped data in this way, classifying by intervals, we are losing some sensitivity and accuracy. For instance, both observations that fell in the interval $35 to $44.99 are counted as having a value of $40, the midpoint.

The weighted mean can also be used when the importance of each observed value is determined by something other than the frequency of occurrence. For instance, a personnel manager may give twice as much importance to an applicant's score on a mechanical aptitude test as he does the score on a manual dexterity test. To find the weighted mean of an applicant's test scores, he would use the appropriate weights as f and the sum of the weights as n in SO&B 3-5, where $\bar{x} = \Sigma fx/n$. If an applicant received a score of 80 on the mechanical aptitude test and a score of 95 on the manual dexterity test, his weighted average would be

$$\bar{x} = \frac{\Sigma fx}{n}$$

$$= \frac{2 \cdot 80 + 1 \cdot 95}{3}$$

$$= \frac{160 + 95}{3}$$

$$= 85$$

THE MEDIAN—GROUPED DATA

We have seen that the median is that value or observation for which half the data are larger than that observation and half the data are smaller. When we are faced with grouped data, the procedure for estimating the median is a little more complex than for ungrouped data. To illustrate the procedure for grouped data, let's reconstruct the data in Table 3-3.

TABLE 3-3 (again).   Money Collected from Machine 204 Per Day

Daily Amount	Frequency (f)	Cumulative Frequency	
$35–$44.99	2	2	
$45–$54.99	5	(F) 7	
(L)$55–$64.99 (i)	(f) 7	14	←Median interval
$65–$74.99	8	22	
$75–$84.99	3	25	
	(n) 25		

Noting that we have a total of 25 observations, the median should be the $(\frac{25}{2})$th or the $(12\frac{1}{2})$th observation. Looking down the cumulative frequency column, we see that the $(12\frac{1}{2})$th observation would be in the class interval $55 to $64.99. We will call this interval the *median interval*. We now need to determine where the median is in this range of $55 to $64.99. We do this by use of the formula

SO&B 3-6: $$\text{Median} = L + i \cdot \frac{\left(\frac{n}{2} - F\right)}{f}$$

where $L =$ lower limit of the median interval
$\quad\quad i =$ length of the median interval
$\quad\quad n =$ total number of observations
$\quad\quad F =$ total number of observations *before* the median interval
$\quad\quad f =$ number of observations in the median interval

In our example we have already identified the median interval as $55 to $64.99. The lower limit of that interval is $55; therefore $L = 55$. The interval is ten units long, so $i = 10$. We have a total of 25 observations, so $n = 25$. Looking at the cumulative frequency column in Table 3-3 (again), we see that there were seven observations $(2 + 5)$ before we got to the median interval, so $F = 7$. The number of observations in the median interval is seven, so $f = 7$ also. Using SO&B 3-6 we obtain

$$\text{Median} = L + i \cdot \frac{\left(\frac{n}{2} - F\right)}{f}$$

$$= 55 + 10 \cdot \frac{\left(\frac{25}{2} - 7\right)}{7}$$

$$= 55 + 7.86$$

$$= 62.86$$

Example 1: Let's look now at a recap of the measures of central tendency and some comparisons of which measure might be the most appropriate. The example we will use involves Mabel Metcher. Ms. Metcher is a secretary at a large corporation and has been taking night classes to improve her skills. In looking for an opportunity to apply her new knowledge, Mabel has become fascinated with the time-sharing teletype computer terminal installed in her office.

She decides to write a program for a computerized dating service and use it on the computer for her friends when the teletype is idle (and the boss is not watching). Her success is greater than she anticipated and soon she has a part-time business going. She collects a fee from her customers based on the degree of satisfaction they experienced from her service. The business now deserves a name, and she appropriately calls her new firm Mabel Metcher's Mate Matcher. She has adopted the slogan, "If you don't have a flame try a match from Mabel's." A sample of Mabel's extracurricular activities is offered in Table 3-4.

TABLE 3-4. Daily Income for Mabel Metcher's Mate Matcher

Day	Income	Day	Income	Day	Income
1	$0	10	$0	19	$0
2	1.30	11	2.50	20	0
3	.25	12	0	21	3.00
4	7.49	13	1.00	22	0
5	0	14	2.00	23	3.75
6	0	15	0	24	0
7	2.50	16	0	25	1.00
8	15.21	17	29.01	26	0
9	4.00	18	.35	27	6.52
					$79.88

Suppose for a moment that these data were gathered by an Internal Revenue Service agent. The IRS is investigating Mabel and, in typical fashion, is not really interested in how she got the money so long as she pays her taxes. The IRS agent would like to estimate Mabel's annual income based on the sample of 27 days. He would, therefore, be interested in calculating the mean. Using SO&B 3-2 the agent calculates

$$\bar{x} = \frac{\Sigma x}{n} = \frac{79.88}{27} = \$2.96$$

Then he estimates her total yearly income from these activities as

$$365 \cdot 2.96 = \$1079.85$$

The IRS will attempt to collect income tax from Mabel for an additional $1079.85.

Suppose the agent wants to condense the data. If so, he can group the data as shown in Table 3-5.

TABLE 3-5. Daily Income for Mabel Metcher's Mate Matcher (Grouped)

Income	Midpoint (x)	Frequency (f)	fx
$0–$ 1.99	1.00	17 (丗 丗 丗 II)	17.00
$2.00–$ 3.99	3.00	5 (丗)	15.00
$4.00–$ 5.99	5.00	1 (I)	5.00
$6.00–$ 7.99	7.00	2 (II)	14.00
$8.00–$ 9.99	9.00	0	0
$10.00–$19.99	15.00	1 (I)	15.00
$20.00–$29.99	25.00	1 (I)	25.00
		27	91.00

Then, calculating the mean by use of SO&B 3-5,

$$\bar{x} = \frac{\Sigma fx}{n} = \frac{91.0}{27} = \$3.37$$

Notice here that the mean of $3.37 using the grouped data is different from the mean of $2.96 he found using the raw or ungrouped data. This difference is caused by the loss of accuracy when he grouped the observations. The smaller the intervals, the more accurate will be the estimated mean from grouped data. The smaller the interval, however, the less summarization is done. Note also that the agent used different sizes of intervals. He did this to keep his summarization or grouped table of data small and still "catch" the two large observations.

If the foregoing analysis had been gathered for another purpose, the median may have been a more appropriate measure than the mean. Suppose Mabel is going to be the keynote speaker at a professional meeting of women in the dating service business, the Miss Matchers. In the keynote address, Mabel will at one point be trying to give the group a general idea of the daily income from her business. For this purpose she may want to use the median instead of the mean. Remember: the median often gives a better *description* of the data than the mean when a few extreme values are present.

Using SO&B 3-3, the median in this case would be found by arranging the data in ascending order and then identifying the $((n + 1)/2)$th observation. This would be the $((27 + 1)/2)$th or the fourteenth observation when the data are arranged in ascending order. The value of that observation is 35 cents. This figure might be more descriptive of the daily amount gained than the mean because the mean was affected by two observations that seem to be relatively high ($15.21 and $29.01).

If Mabel had grouped the observations as the IRS agent had done, the median would be found in a different manner. The data appeared in grouped form as

Income	Frequency	
$0–$ 1.99	17	← Median interval
$2.00–$ 3.99	5	
$4.00–$ 5.99	1	
$6.00–$ 7.99	2	
$8.00–$ 9.99	0	
$10.00–$19.99	1	
$20.00–$29.99	1	
	27	

Using SO&B 3-6, Mabel would calculate

$$\text{Median} = L + i \cdot \frac{\left(\frac{n}{2} - F\right)}{f}$$

$$= 0 + 2.00 \cdot \frac{(13.5 - 0)}{17}$$

$$= 0 + 1.59$$

$$= \$1.59$$

Again note that grouping data causes some loss of accuracy. The median estimated by the grouped data is $1.59 whereas the actual median from the ungrouped data was $0.35.

The third measure of central tendency, the mode, may have been the most useful in a different situation. Suppose that one of Mabel's acquaintances is considering going into the computerized dating business for extra income also. This acquaintance is very cautious and is not one to put a lot of effort into something unless the rewards are guaranteed to be high. She may look at the sample data and note the mode of zero (the observation 0 occurs 12 times). She may then choose not to go into the business because "most of the time" the reward is zero. The mode may also be of value to Mabel in her fight with the IRS. The mode of zero would be better for her tax status than the mean of $2.96, but it's doubtful whether the IRS would go for that!

AUTOPSY

So far we have concentrated pretty much on a discussion of which measure of central tendency can be used and how it is calculated. It is just as important, though, to be able to analyze these measures when you have received them yourself.

A purchasing agent for a toy manufacturer may be considering three different brands of batteries to be used in toys. Suppose a report from the testing laboratory shows the data presented in Table 3-6.

By looking at the first brand, Wink, the purchasing agent should realize that the lifetime of these batteries is perfectly symmetric. In

TABLE 3-6. Lifetime of Batteries (in hours)

	Brand of Battery		
	Wink	Ace	Tops
Mean lifetime	3.5	3.5	3.5
Median lifetime	3.5	2.0	5.0

other words, if we graphed the frequency of occurrence of the life-
times of a group of these batteries it would look like Figure 3–2.

The data on the second brand, Ace, indicate that most of these
batteries wear out very fast but a few have a high life expectancy.
When the median is less than the mean, the numbers probably "tail
off" (are *skewed*) to the right. This tendency is illustrated in Figure 3–3.

FIGURE 3–2

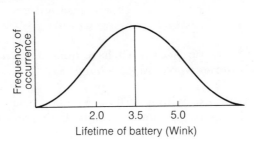

Lifetime of battery (Wink)

FIGURE 3–3

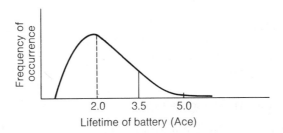

Lifetime of battery (Ace)

The data on the third brand, Tops, suggest that a few batteries
wear out almost immediately but a lot of them last a long time. When
the median is higher than the mean, the numbers or data probably
"tail off" (are *skewed*) to the left. This tendency is illustrated in Figure
3–4.

The purchasing agent may then assess his position as follows
(assuming prices are equal on all three brands):

> If we use Tops, there will be a few very irate customers
> whose batteries will fail almost instantly.

FIGURE 3-4

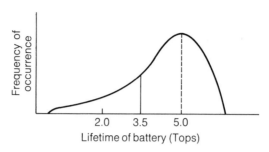

If we use Aces, we'll have quite a few customers whose batteries won't last very long.

If we use Winks, we'll have a good balance between some lasting a short time and some lasting an unduly long time.

He then astutely decides to buy the Winks because his boss's son-in-law is the battery salesman for that company.

SUMMARY

Important Terms

Mean: the arithmetic average

Median: the halfway point of a group of data; cuts the data in half

Mode: observation that occurs most often

Parameter: a characteristic of a population

Population: the entire group or set with which we are concerned

Sample: a subgroup or subset of a population

Statistic: a characteristic of a sample

Mean for Ungrouped Data

SO&B 3-2: $$\bar{x} = \frac{\Sigma x}{n}$$

Mean for Grouped Data

SO&B 3-5: $$\bar{x} = \frac{\Sigma fx}{n}$$

Median for Odd Number of Ungrouped Observations

SO&B 3-3:   Median = value of the $\left(\frac{n+1}{2}\right)$th observation

Median for Even Number of Ungrouped Observations

SO&B 3-4:   Median = $\dfrac{\text{value of } (n/2)\text{th observation} + \text{value of } (n/2 + 1)\text{th observation}}{2}$

Median for Grouped Data

SO&B 3-6: $$\text{Median} = L + i \cdot \frac{\left(\frac{n}{2} - F\right)}{f}$$

PROBLEMS

1. The ages of the five people in an office are 19, 27, 32, 45, 45. Find the mean, median, and mode of these ages.

2. The salesmen for an automobile dealership have the following number of new car sales for the month of July:

Salesman	Sales
Pete	12
George	8
Sam	28
Bill	12
Frank	13
Marty	10
David	12
Moose	11
Rod	12
Lefty	15

Find the mean, median, and mode of the salesmen's results.

3.   In Problem 2, suppose the sales manager is hiring a new salesman. What would be the most realistic estimate of average sales he could give the new salesman? Could he realistically use the July data as an estimate for all months?

4.   Fred's Fertilizer Company has sold ten 100-pound bags of fertilizer, twenty 50-pound bags, fifteen 25-pound bags, and four 1-pound bags. Find the average weight of their sales.

5.   A manufacturing company is analyzing the repair time needed for breakdowns on a certain machine. Their records show the following data:

Hours Needed for Repair	Frequency
0 and below 4	0
4 and below 8	5
8 and below 12	3
12 and below 16	4
16 and below 20	1

Find the mean of the repair time.

6.   A sample of a retail store's weekly sales records shows the following:

Weekly Sales	No. of Weeks
$1501–$2500	3
$2501–$3500	5
$3501–$4500	1
$4501–$5500	4
	15

Find the mean and the median of weekly sales.

7. A sample of eight days revealed the following observations on the number of cars illegally parked in a restricted area:

No. of Cars
4
0
6
2
2
5
2
7

Find the mean and the median.

8. The lifetime of a certain brand of tires has a mean of 20,000 miles and a median of 14,000 miles. Describe the performance of these tires.

9. You want to give a fair appraisal of the average salary of a company's employees. Should you use the mean, the median, or the mode? Why?

10. The assistant comptroller of a company has gathered the following data on the average time spent on coffee breaks by employees: mean 14 minutes; median 10 minutes. Describe the coffee break habits of employees.

11. Make up an example, from the real world, of a population parameter and a corresponding sample statistic.

SELECTED REFERENCES

Guenther, William C. *Concepts of Statistical Inference.* Second edition. New York: McGraw-Hill, 1973. Especially Chapter 3.

Miller, Irwin and John E. Freund. *Probability and Statistics for Engineers.* Englewood Cliffs, N.J.: Prentice-Hall, 1965. Especially Chapter 6.

Spurr, William A. and Charles P. Bonini. *Statistical Analysis for Business Decisions.* Revised edition. Homewood, Ill.: Irwin, 1973. Especially Chapter 3.

Tanur, Judith M., Frederick Mosteller, William H. Kruskal, Richard F. Link, Richard S. Pieters, and Gerald R. Rising. *Statistics: A Guide to the Unknown.* San Francisco: Holden-Day, 1972.

Wallis, W. A. and H. V. Roberts. *The Nature of Statistics.* New York: Free Press, 1962 Especially Chapters 6, 7.

Yamane, Taro. *Statistics: An Introductory Analysis.* Third edition. New York: Harper & Row, 1973. Especially Chapter 3.

APPENDIX
USING CODED DATA TO CALCULATE A MEAN

IN MY WORK I have always had some type of calculator or computer available. In many academic situations, however, a would-be analyst must do the calculations by hand. If this is the case, this appendix describes a shortcut method for finding the mean of large, cumbersome numbers. The shortcut involves coding the original data by subtracting a constant from each observed value or dividing each observed value by a constant, or both. To illustrate the technique, suppose we have taken a sample of five students and asked each how many hours he spent studying statistics during the course. The results (in hours) may be as follows:

x
1007
1005
1004
1006
1003
5025

(You may think these estimates are a little high, but it has been sworn to me through tear-soaked handkerchiefs that they are accurate.)

To find the average amount of time spent studying statistics, we would calculate

$$\bar{x} = \frac{\Sigma x}{n}$$

$$= \frac{5025}{5}$$

$$= 1005$$

If we found these figures cumbersome to manipulate, we could code the data by subtracting a constant A, from each observation. If we let $A = 1000$, we would have

$x - A$
7
5
4
6
3
25

The mean of the coded data is

$$\bar{x} \text{ (coded)} = \frac{25}{5}$$

$$= 5$$

These numbers are considerably easier to handle when you are doing the calculations by hand. To convert the coded mean back to the original data, we merely *add* the constant A to the coded mean:

$$\bar{x} = \bar{x} \text{ (coded)} + A$$

$$= 5 + 1000$$

$$= 1005$$

which is the same answer we got before.

We can also code the data by dividing each observation by a constant K. Suppose that in another school a similar sample of five students reported the following hours spent studying:

x
900
5000
3000
1000
100
10,000

These data could be coded by dividing each observation by $K = 1000$. The coded data would appear as

x/K
.9
5.0
3.0
1.0
.1
10.0

Finding the mean of the coded data, we would calculate

$$\bar{x} \text{ (coded)} = \frac{10.0}{5}$$

$$= 2.0$$

To convert this coded mean back to the original data, we merely *multiply* the constant K by the coded mean:

$$\bar{x} = \bar{x} \text{ (coded)} \cdot K$$

$$= 2.0 \cdot 1000$$

$$= 2000$$

which is the mean of our original data.

If we code an original set of data by first subtracting a constant A

from each observation and then divide each result by a constant K, the mean of the original data is found by

$$\bar{x} = [\bar{x}\,(\text{coded}) \cdot K] + A$$

This, of course, tells us to multiply the resulting coded mean by K and *then* add the constant A to this product.

CHAPTER 4
DESCRIBING DISPERSION

OBJECTIVES: After studying this chapter and working the problems, you should be able to:

Explain why knowing a central tendency measure such as the mean may not tell you enough about the data.

Explain the concept of range and demonstrate how extreme values can affect this measure.

Develop your own working definition of dispersion of data.

Define standard deviation and calculate this value for population and sample data (grouped and ungrouped).

Explain the significance of the coefficient of variation.

Explain why data from two different sources must sometimes be standardized.

Explain the meaning and use of quantiles.

IN THE PREVIOUS chapter we learned about describing the central tendency of data. A measure of central tendency generally does not tell us enough information, though. We need something more. This can be readily seen by visualizing a restaurant owner who must schedule his waitresses. Suppose he knows that he has an average of 70 customers per hour. Does that mean he should schedule his waitresses so that there are always enough waitresses to handle about 70 customers per hour? Looking at the fluctuations in a restaurant business with large crowds at lunch and supper time and very slow

periods between, one would have to conclude that scheduling waitresses to handle an average of 70 customers would be disastrous.
During the peak hours you would have irate customers waiting for
service, and during slack times you would have waitresses standing
around with nothing to do. The mean of 70 customers leaves something to be desired in describing the number of customers per hour.

To understand the void left by knowing only the mean or some
measure of central tendency, consider the plight of Marcia and
John—two singles who have gone to Mabel Metcher's Mate Matcher
to try their luck at meeting new dates. Both of them are concerned
with physical beauty, so they ask the computer to give them only the
body dimensions of potential dates. The computer gives each of them
two leads. Male A and male B, potential dates for Marcia, both have a
mean body diameter of 36 inches. Female C and female D, potential
dates for John, both have a mean body diameter of 30 inches. Which
dates should they choose? Again the mean doesn't give them enough
information. Figure 4–1 shows the two men and the two women the
computer has in mind as dates for Marcia and John.

The measurement we need to describe the data in the preceding
examples more completely is some measure of variation. We need

FIGURE 4–1

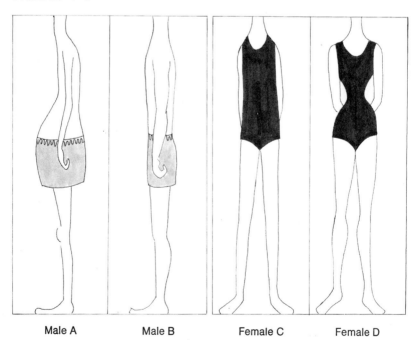

Male A   Male B   Female C   Female D

a measure of how scattered, or how compact, or how dense, or how homogeneous the data are. The restaurant owner needs to measure how scattered the customers per hour data are: this measure would tell him that he does not consistently have about 70 customers each hour. The two single daters need to know how much variation exists in the body diameters of their potential dates. John would probably not be interested in female C, who measures 30 inches in diameter everywhere from her shoulders to her feet, but female D shows a lot of variation or scatter even though her average diameter is the same as that of female C.

The simplest measure of dispersion or variation is the *range*. The range is found by subtracting the smallest observation from the largest observation:

SO&B 4-1:   Range = maximum − minimum

In the example of Chapter 3 concerning the amount of purchase in the Veni, Vidi, Vendo vending machine, we had

Customer (i)	Amount of Purchase
1	5¢
2	10
3	5
4	25
5	5
6	5
7	−45
8	5
9	10
10	5

And if we were looking for the range of the amount of purchase, we would calculate from SO&B 4-1:

$$\text{Range} = 25 - (-45)$$
$$= 70¢$$

The range would be a useful statistic for our restaurant owner: it would give him an indication of how much the hourly customer traffic varies.

Although the range is a simple and easy to understand measurement, its simplicity limits its general usefulness. It is obviously sensitive to rare but unusually large or small observations. This sensitivity shows up in the observation of —45 cents caused by the little old lady kicking the machine. I believe it is safe to assume that there aren't too many little old ladies going around kicking vending machines and getting money back, yet this observation had an extreme effect on the range. Hence even if the observation of —45 cents happens only once every five years it has still greatly affected our range.

Remember what we are trying to measure. Put in a number of different ways that say basically the same thing, we are trying to measure:

The dispersion of the data

The density of the data

How closely packed the data are

How homogeneous the data are

How varied the data are

Take your pick of any of these—whichever means the most to you.

One measure of dispersion, which has an intuitive meaning, is the *average deviation*. We can visualize finding the distance between each observation and the mean. The average of these distances would, rationally, give us a measure of the dispersion or scatter of the data. We could find this average deviation by using the formula

$$\text{Average deviation} = \frac{\Sigma\,|x - \bar{x}|}{n}$$

where the verticals (| |) mean "the absolute value of" or "treat all negative numbers as positive numbers." Each $(x - \bar{x})$ is the distance or deviation of the observation from the mean. And summing these deviations and dividing by the number of observations n yields an "average" deviation. Although the average deviation would give us an intuitive measure of dispersion, it is seldom used because we can do nothing further with it mathematically.

The most commonly used measure of dispersion is the *standard deviation*. The standard deviation is more complex and less intuitive than the average deviation, but it has mathematical properties that make it much more useful for further analytical work, as we will see in future chapters.

THE STANDARD DEVIATION

Like the average deviation, the standard deviation also measures distances from each observation to the mean. Instead of taking the absolute value of this distance, though, it squares that distance and then later attempts to "unsquare" it by taking the square root of the average of the sum of these squared differences. We will look at the standard deviation in three different situations: the population standard deviation with ungrouped data, the sample standard deviation with ungrouped data, and the sample standard deviation with grouped data.

THE POPULATION STANDARD DEVIATION (σ) — UNGROUPED DATA

If we have information on *all* the group with which we are concerned, then we are looking for the standard deviation of a population. We will use the Greek symbol σ to designate a population standard deviation. First we will look at a *definitional* formula. Later in this chapter we will look at a *computational* formula for the standard deviation. The computational formula is much easier to use, but the definitional formula more clearly illustrates what is actually happening. The definitional formula is

SO&B 4-2:
$$\sigma = \sqrt{\frac{\Sigma(x - \mu)^2}{N}}$$

where x = individual observation
μ = population mean
N = number in the population

Step by step, this formula (1) finds how far each observation is from the mean $(x - \mu)$; then (2) it squares each of these differences or dispersions. The next step (3) is finding the sum of these squared differences $[\Sigma(x - \mu)^2]$, after which (4) we divide this total by the number of observations $[\Sigma(x - \mu)^2/N]$ — which, in effect, is an average squared difference from the mean. To get this back into the right units, in other words to "unsquare" it, the final step (5) is to take the square root:

$$\sqrt{\frac{\Sigma(x - \mu)^2}{N}}$$

To demonstrate the use of this formula, let's take the data from the Veni, Vidi, Vendo vending machine and slightly redefine them. Suppose that the machine used for the observations was in operation for only one hour and that the ten observations were *all* the customers during that hour. Our data now represent a population (the entire group or set with which we are concerned). We have already found in the previous chapter that the mean μ is 3 cents. We would find σ as shown in Table 4-1.

TABLE 4-1. Population Standard Deviation for
Veni, Vidi, Vendo Company

Customer (i)	Amount of Purchase	$x - \mu$ ($\mu = 3$)	$(x - \mu)^2$
1	5¢	2	4
2	10	7	49
3	5	2	4
4	25	22	484
5	5	2	4
6	5	2	4
7	−45	−48	2304
8	5	2	4
9	10	7	49
10	5	2	4
	30¢		2910

$$\sigma = \sqrt{\frac{\Sigma(x - \mu)^2}{N}} = \sqrt{\frac{2910}{10}} = \sqrt{291.0} = 17.06$$

In step 1 we found how far each observation is from the mean (column 3). In step 2 we squared each of these differences (column 4) and then in step 3 we found the sum of these differences squared (total of column 4). We then performed step 4 by dividing the sum by N and finally did step 5 by finding the square root of the answer from step 4. The population standard deviation σ is 17.06 for these data. The square of the standard deviation, σ^2 for the population, is called the *variance*. The variance is sometimes used as a measure of dispersion also. For the example we just worked, the variance σ^2 would be 291.

THE SAMPLE STANDARD DEVIATION (S)—UNGROUPED DATA

If we have information on only a part or a subset of the group with which we are concerned, then we are looking for the standard deviation of a *sample*. When we are calculating the standard deviation for a sample S, we make two changes in the formula for the standard deviation of a population σ. The formula for a sample standard deviation is

SO&B 4-3:
$$S = \sqrt{\frac{\Sigma(x - \bar{x})^2}{n - 1}}$$

Note that instead of finding the differences of each observation from μ, we are finding the differences of the observations from $\bar{x}$, the sample mean. Note also that instead of dividing by N, we are dividing by $n - 1$. The fact that we divide by $n-1$ instead of n appears to bother many beginning students. It seems logical to believe that you would divide by n. The reason for dividing by $n-1$ can be proved at length mathematically, but a brief explanation can be given here.

One of the basic uses of sample statistics is for estimating population parameters. If, in the long run, a statistic will average out to the value of the population parameter, then we say the statistic is an *unbiased* estimate of its respective parameter. The sample statistic $\bar{x}$ is an unbiased estimate of the population parameter μ in this respect. But if we use n instead of $n - 1$ to calculate the value of S^2, then the average of S^2 will not equal σ^2 in the long run. The difference can be compensated by using $n - 1$ in the denominator of the formula for S^2 and, hence, S.

We can demonstrate the use of this formula by returning to the data from the Veni, Vidi, Vendo Company. We will redefine the data now to be a sample of customers by saying that the ten observations represent a sample taken from many customers of this particular vending machine. We would then compute the sample standard deviation as shown in Table 4-2.

The standard deviation for these *sample* data is 17.98. The variance S^2 is 323.33. Note that $\Sigma(x - \bar{x})$, the total of the third column, is zero. This result is a must. If $\Sigma(x - \bar{x})$ is not equal to zero you have made an arithmetic error. Note also that the use of S to denote the standard deviation of a sample is common notation, as is S^2 to denote the variance of a sample.

TABLE 4-2.   Sample Standard Deviation for Veni, Vidi, Vendo Company

Customer (i)	Amount of Purchase (x)	$x - \bar{x}$ ($\bar{x} = 3$)	$(x - \bar{x})^2$
1	5¢	2	4
2	10	7	49
3	5	2	4
4	25	22	484
5	5	2	4
6	5	2	4
7	−45	−48	2304
8	5	2	4
9	10	7	49
10	5	2	4
	30¢	0	2910

$$S = \sqrt{\frac{\Sigma(x - \bar{x})^2}{n-1}} = \sqrt{\frac{2910}{9}} = \sqrt{323.33} = 17.98$$

THE SAMPLE STANDARD DEVIATION (S) – GROUPED DATA

Instead of having the data listed individually as in Table 4-2, we are sometimes presented with data in grouped form. If we group the data from Table 4-2 concerning the Veni, Vidi, Vendo Company, we will have what appears in Table 4-3.

The definitional formula for grouped sample data is

SO&B 4-4:
$$S = \sqrt{\frac{\Sigma f(x - \bar{x})^2}{n-1}}$$

Applying this formula to the grouped data as shown in Table 4-3, we see that $S = 17.98$ and $S^2 = 323.33$ – which is the same result we got from Table 4-2 when we used the data ungrouped. Of course, we'd better get the same result since the data are the same in both cases, only in different form.

TABLE 4-3. Sample Standard Deviation for
Veni, Vidi, Vendo Company (Grouped Data)

Amount of Purchase (x)	Frequency (f)	$x - \bar{x}$ ($x = 3$)	$(x - \bar{x})^2$	$f(x - \bar{x})^2$
−45¢	1	−48	2304	2304
5	6	2	4	24
10	2	7	49	98
25	1	22	484	484
				2910

$$S = \sqrt{\frac{\Sigma f(x - \bar{x})^2}{n - 1}} = \sqrt{\frac{2910}{9}} = \sqrt{323.33} = 17.98$$

THE COMPUTATIONAL FORMULAS FOR STANDARD DEVIATION

We have spent enough time now with the definitional formulas for standard deviations (SO&B 4-2, 4-3, 4-4). You should now recognize that a standard deviation has, as a basic ingredient, a measure of the distance of *each* observation from the mean value. The computational formulas we will be using are in a form that is much easier to use than the definitional formulas. However, we began with the definitional formulas because it is easier to discern from them the basic idea of a variance and standard deviation.

The computational formulas you should generally use are the following. Standard deviation of a population:

SO&B 4-5: $$\sigma = \sqrt{\frac{N\Sigma x^2 - (\Sigma x)^2}{N^2}}$$

Standard deviation of a sample with ungrouped data:

SO&B 4-6: $$S = \sqrt{\frac{n\Sigma x^2 - (\Sigma x)^2}{n(n - 1)}}$$

Standard deviation of a sample with grouped data:

SO&B 4-7:

$$S = \sqrt{\frac{n\Sigma fx^2 - (\Sigma fx)^2}{n(n-1)}}$$

These computational formulas are only an algebraic manipulation of the corresponding definitional formulas and give exactly the same results. They are much easier to use, though, because you don't have to find the mean before you start and most calculators can accumulate both the sums of the x's and the sum of the x^2's in one pass. You should refer to the instruction manual of whatever calculator you may have available to verify this last statement.

Example 1: Returning to our friend Mabel Metcher, we can now use our knowledge of measures of dispersion to analyze the data gathered on her activities. In Chapter 3 we had a sample of the income of her activities as shown in Table 4-4.

In this table we have added a column of x^2 to the original Table 3-4 showing the observation number and the observation value. We want to compute the standard devia-

TABLE 4.4   Standard Deviation for Mabel's
Daily Income

Day	Income (x)	x^2	Day	Income (x)	x^2
1	$0	0	14	$2.00	4.0000
2	1.30	1.6900	15	0	0
3	.25	.0625	16	0	0
4	7.49	56.1001	17	29.01	841.5801
5	0	0	18	.35	.1225
6	0	0	19	0	0
7	2.50	6.2500	20	0	0
8	15.21	231.3441	21	3.00	9.0000
9	4.00	16.0000	22	0	0
10	0	0	23	3.75	14.0625
11	2.50	6.2500	24	0	0
12	0	0	25	1.00	1.0000
13	1.00	1.0000	26	0	0
			27	6.52	42.5104
				$79.88	1230.9722

tion of a sample using ungrouped data; hence we will use SO&B 4-6. In SO&B 4-6,

$$S = \sqrt{\frac{n\Sigma x^2 - (\Sigma x)^2}{n(n-1)}}$$

and we need the sum of each of the x's squared, Σx^2, and the square of the total of the x's, $(\Sigma x)^2$. In Table 4-4 we have the x^2 column of each observation. The sum of this column is 1230.9722. Therefore

$$S = \sqrt{\frac{27(1230.9722) - (79.88)^2}{27(26)}}$$

$$= \sqrt{\frac{26855.4350}{702}}$$

$$= \sqrt{38.2556}$$

$$= 6.19$$

Including this value with our previous calculation of the mean, $2.96, we would describe Mabel's daily extracurricular income as having a mean $\bar{x}$ of $2.96 and a standard deviation S of $6.19. If you were viewing these two summary statistics without having seen the original data, you would know that the data "center" around a little less

TABLE 4-5. Standard Deviation (Grouped Data) for Mabel's Daily Income

Income ($)	Midpoint (x)	Frequency (f)	x^2	fx^2	fx
0–$ 1.99	1.00	17	1	17	17
$2.00–$ 3.99	3.00	5	9	45	15
$4.00–$ 5.99	5.00	1	25	25	5
$6.00–$ 7.99	7.00	2	49	98	14
$8.00–$ 9.99	9.00	0	81	0	0
$10.00–$19.99	15.00	1	225	225	15
$20.00–$29.99	25.00	1	625	625	25
		27		1035	91

than \$3 and have a considerable amount of scatter, as
indicated by $S = 6.19$.

If the IRS agent had chosen to condense the raw data
as he did in Chapter 3, then we would have the data
grouped as in Table 4-5.

Once again we have inserted additional columns in the
original Table 3-5. This time we have added a column for
x^2 and fx^2. Using SO&B 4-6 for a sample standard deviation from grouped data, we have

$$S = \sqrt{\frac{n\Sigma fx^2 - (\Sigma fx)^2}{n(n-1)}}$$

$$= \sqrt{\frac{27(1035) - (91)^2}{27(26)}}$$

$$= \sqrt{\frac{19{,}664}{702}}$$

$$= \sqrt{28.01}$$

$$= 5.29$$

The \$5.29 for S with grouped data is slightly different
from the \$6.19 for S with ungrouped data. We are re-
minded once again that grouping data into intervals causes
some loss of accuracy, but this loss is generally more than
made up for in the gain in ease of working with the smaller,
summarized grouped data.

The columns concerning x^2 have been added in these
tables of data just to show you how to work with them.
With most calculators it is not necessary to record each
x^2 and fx^2 and, as a matter of fact, this task becomes a
time-consuming chore. (See the appendix at the end of
this chapter for a shortcut method of calculating by hand.)

COEFFICIENT OF VARIATION

Most beginning students in statistics have trouble interpreting the
standard deviation in terms of "what does it tell me?" First of all, the
standard deviation as a measure of variation can be compared to the
standard deviation of other groups of data. The analyst can then

make relative comparisons, such as "these data are more scattered than another group of data."

Interpreting the standard deviation directly as a descriptive statistic is not so direct a procedure because the standard deviation is a relative measure with respect to the data being described. For instance, if we have a standard deviation of $S = 4$ for data that average 1 million, we would probably think that these data do not vary much. If we had a standard deviation of $S = 5$ for data that averaged 5, however, we would probably think that these data vary a lot. It takes practice with standard deviation—calculating it from data that can be seen—before the analyst can develop a true feel of what the standard deviation is saying in terms of dispersion or variation.

One measurement that is used to relate the standard deviation to the mean is the *coefficient of variation*. The coefficient of variation (V) is found by

SO&B 4-8:
$$V = \frac{S}{\bar{x}}$$

standard of deviation / mean

In the ungrouped data for Mabel's activity we found the mean ($\bar{x} = \$2.96$) and the standard deviation ($S = \$6.19$). Putting these values into a coefficient of variation using SO&B 4-8, we have

$$V = \frac{S}{\bar{x}}$$

$$= \frac{6.19}{2.96}$$

$$= 2.09$$

A coefficient of variation of 2.09 tells us that the standard deviation of these data is more than twice as high as the mean—in other words, it is highly scattered. The coefficient of variation has an interpretation that is relative to the situation. In a life or death situation (producing medicine, for instance) one would conceivably want a coefficient of variation of less than .01.

STANDARDIZING DATA TO MAKE COMPARISONS

Thus far in our discussion we have used the standard deviation strictly as a way of describing data. There is another valuable use of the standard deviation, though: it can be used in making value judgments about different groups of data. This application can be illustrated by imagining yourself in a testing situation. Suppose you receive a score of 120 on an aptitude test in accounting while on an aptitude test in computer programming you receive a score of 110. Both tests have an average score of 100. Which did you do better on? You cannot automatically say "accounting with 120" without considering other factors. Perhaps a number of people have received better than 120 on the accounting test but no one else has ever received a score as high as 110 on the computer programming exam. In other words, it might be just as hard or harder to get 110 on the programming test as it is to get 120 on the accounting test—even though the average score of all test takers is 100 on both tests.

In statistical terms, the standard deviation of the scores on the accounting test may be much higher than the standard deviation of the scores on the computer programming test. To truly evaluate how well we did, in comparing our scores we would have to take the standard deviation into consideration. If the standard deviation of the scores on the accounting test is 20, then we scored *one* standard deviation above the average of 100. If the standard deviation on the computer programming exam scores is 5, then we scored *two* standard deviations above the mean of 100. Our performance on the programming exam was actually better than our performance on the accounting exam.

Consider a situation in which you, the analyst or decision maker, want to measure the performance of a group of district sales managers. To simplify the situation, there are only two products being sold. Let's return to an earlier example from Chapter 2 and say we are now marketing two different styles of yard lights—one that looks like a dandelion and one that looks like a droopy daffodil. These two products sell equally well with an average monthly sales volume of 500 units for each product in each of our ten districts. Each district has a district sales manager (DSM), and we are trying to rank these ten DSMs from 1 through 10 based on their sales performance.

A common manner of rating the DSMs would be to add the two sales volumes together (number of dandelions sold and number of droopy daffodils sold) for each DSM and rank them on the sum. This approach is in many cases deceptive, even though the ranking system has the advantage of being simple. The deceptiveness comes in the fact that even though the mean of the two products is the same, the standard deviations may be considerably different.

Table 4-6 shows one month's performance in the two product

lines by the ten districts. In this table we see that the average of the
two products is 500 for each. But the standard deviation for dandelions, S_1, is 50 whereas the standard deviation for daffodils, S_2, is 5.
Now look at the ranking of the DSMs based on their total sales. You
will see that it is the same as the ranking they would receive if they
were being rated on sales of dandelion lights alone.

TABLE 4-6.   Sales for One Month by District
by Product

District	Type of Light Dandelions	Daffodils	Total	Rank Based on Total
1	600	491	1091	1
2	530	506	1036	2
3	525	503	1028	3
4	520	502	1022	4
5	500	500	1000	5
6	500	498	998	6
7	490	499	989	7
8	470	498	968	8
9	447	495	942	9
10	418	508	926	10
$\bar{x}_1 = 500$	$\bar{x}_2 = 500$	$S_1 = 50.0$	$S_2 = 5.0$	

The products are supposed to be equally weighted, yet why is district
1 ranked number 1 even though they were *last* in droopy daffodil
sales? By the same token, why is district 10 ranked last even though
they were *first* in droopy daffodil sales? The answer lies in the standard
deviations. We have already observed that the dandelion design has a
much higher standard deviation in its sales than does the droopy
daffodil design. Look at Table 4-6 to see what this does to the ranking.
The first-place district has gained 70 points ($600 - 530$) over the
second-place district in dandelion design sales. There is no way that
even superb droopy daffodil sales are likely to overcome this point
spread because the droopy daffodil sales are typically closely packed.
In general the high dispersion of the dandelion design sales leads to
large point spreads, while the low dispersion of the droopy daffodil
sales results in only small point spreads even though it was probably
just as difficult to obtain droopy daffodil sales of 508 as it was to
obtain dandelion sales of 600.

If the analyst truly wants the two products weighted equally, he must *standardize* the data. In our example we can standardize the data by dividing each element by the appropriate standard deviation. The analyst must keep in mind that when he is adding items together to develop a rating or rank, the magnitude of each item (measured by its mean) does *not* affect the final ranking. It is the standard deviation or dispersion that determines the item's weight or effect on the rank. If we standardize the data in Table 4-6, we will obtain the ranking shown in Table 4-7.

In Table 4-7 we developed the standardized column for the dandelion design by dividing each of the original unit sales by the standard deviation of 50. We developed the standardized column for the droopy daffodil design by dividing each of the original unit sales

TABLE 4-7. Ranking of Districts for One Month:
Original Versus Standardized Data

District	Dandelions Original	Dandelions Standardized	Daffodils Original	Daffodils Standardized	Total Original	Total Standardized
1	600	12.0	491	98.2	1091	110.2
2	530	10.6	506	101.2	1036	111.8
3	525	10.5	503	100.6	1028	111.1
4	520	10.4	502	100.4	1022	110.8
5	500	10.0	500	100.0	1000	110.0
6	500	10.0	498	99.6	998	109.6
7	490	9.8	499	99.8	989	109.6
8	470	9.4	498	99.6	968	109.0
9	447	8.9	495	99.0	942	107.9
10	418	8.4	508	101.6	926	110.0

Rank Based on Total

District	Original	Standardized	District	Original	Standardized
1	1	4	6	6	7
2	2	1	7	7	7
3	3	2	8	8	9
4	4	3	9	9	10
5	5	5	10	10	5

by the standard deviation of 5. Notice that when the two products are weighted equally by standardizing the items to be added, the ranks change considerably.

If the decision maker *wants* the elements to be weighted unequally, he can control the desired weights through standardization. If the droopy daffodil design has more profit per sale than the dandelion design, for instance, the decision maker may want it to be weighted twice as heavily. He would do this by standardizing both items with the technique just demonstrated. Then he would multiply the standardized droopy daffodil sales by 2 before adding them to the dandelion sales. This procedure would result in a final ranking that is truly twice as affected by the droopy daffodil design as by the dandelion.

QUANTILES—A MEASURE OF PLACEMENT

We sometimes find a descriptive statistic called a *quantile*—a fractional measure of position. The quantile is a value that is equal to or exceeds a fraction of the data or observations. The most common quantiles are *quartiles,* which divide data into quarters, and *percentiles,* which divide data into hundredths.

The first quartile, Q_1, is equal to or exceeds 25 percent of the data. The second quartile, Q_2, equals or exceeds 50 percent of the data. The third quartile, Q_3, equals or exceeds 75 percent of the data. The fourth quartile, Q_4, is of course the largest observation since it equals or exceeds 100 percent of the data.

Percentiles are often used to designate performance on standardized tests. For instance, if you are in the 96th percentile, P_{96}, your score is equal to or exceeds 96 percent of the scores on that test.

Notice that the second quartile, Q_2, and the 50th percentile, P_{50}, are the median value. Quantiles can be determined in a fashion similar to the way we determined the median.

AUTOPSY

At this point you should know how to calculate a standard deviation, although you probably will not sense exactly what it tells you until you have worked with it for a while. We will be using the standard deviation extensively in later chapters: it is one of the major concepts used in statistical analysis. We have also been introduced to some fairly standard notation in this chapter. Table 4-8 updates this notation from Table 3-1.

TABLE 4-8.   Statistical Notation

Item	For a Population	For a Sample
Number of items	N	n
Mean	μ	$\bar{x}$
Variance	σ^2	S^2
Standard deviation	σ	S

SUMMARY OF FORMULAS

Range

SO&B 4-1: $\text{Range} = \text{maximum} - \text{minimum}$

Standard Deviation of a Population

SO&B 4-5: $$\sigma = \sqrt{\frac{N\Sigma x^2 - (\Sigma x)^2}{N^2}}$$

Standard Deviation of a Sample with Ungrouped Data

SO&B 4-6: $$S = \sqrt{\frac{n\Sigma x^2 - (\Sigma x)^2}{n(n-1)}}$$

Standard Deviation of a Sample with Grouped Data

SO&B 4-7: $$S = \sqrt{\frac{n\Sigma fx^2 - (\Sigma fx)^2}{n(n-1)}}$$

Coefficient of Variation

SO&B 4-8: $$V = \frac{S}{\bar{x}}$$

PROBLEMS

1. As the recently appointed manager of a water quality testing facility, you are concerned about the capacity of the equipment necessary for testing. An assistant tells you that the equipment currently in use can analyze an average of ten water samples per day.

Would you approve a machine that could analyze twelve water samples a day? Would the problem become more acute if the government required the samples to be taken and analyzed on the last working day of the month?

2. Using SO&B 4-5, verify that the computational formula gives the same result as the definitional formula for the data in Table 4-2.

3. In Table 4-6, verify that the standard deviation of the dandelion sales is 50.0 and the standard deviation of the droopy daffodil sales is 5.0.

4. Find the range and standard deviation of the ages in Problem 1 of Chapter 3.

5. Find the range, standard deviation, and coefficient of variation for the data in Problem 2 of Chapter 3.

6. In Problem 5 above, eliminate Sam's sales of 28 units from the data and find the range and standard deviation of the remaining nine observations. Compare your answers with those in Problem 5 and comment on the effect of the extreme value on the range and the standard deviation.

7. Find the standard deviation of the weight of sales in Problem 4 of Chapter 3.

8. Find the standard deviation and coefficient of variation for the data in Problem 5 of Chapter 3.

9. Find the standard deviation of the weekly sales shown in Problem 6 of Chapter 3.

10. A large trucking firm has a policy of changing the tires on their trucks at 20,000 miles regardless of their condition. They are testing two new brands of tires and have collected the following information:

	Brand A	Brand B
Average lifetime (miles)	21,000	21,000
Median lifetime (miles)	21,000	19,000
SD of lifetime (miles)	1,000	7,000

If you were the purchasing agent for the firm, which of the two brands would you recommend, assuming comparable prices and services?

11.   A personnel manager wants to rank six applicants for a job. He has their scores on a general aptitude test and the interviewer's opinion of each. This opinion is a number between 0 and 100, where 100 is the highest possible rating. He would like to combine these two scores and rank the applicants according to their total. He would also like to have the two scores exert an *equal* effect on the final rank. The scores are as follows:

Applicant	Test Score	Interviewer's Opinion
1	80	50
2	85	100
3	79	85
4	83	80
5	74	75
6	79	90

Make the ranking for the personnel manager by standardizing the scores for each test.

12.   Prove that

$$\Sigma \frac{(x - \bar{x})^2}{n - 1} = \frac{n\Sigma x^2 - (\Sigma x)^2}{n(n - 1)}$$

13.   A hospital is checking the number of days spent in the hospital for pneumonia. A sample shows the following observations:

Patient	No. of Days in Hospital
1	7
2	5
3	6
4	5
5	3
6	4
7	5
8	3

Find the mean and the standard deviation of the number of days spent in the hospital for pneumonia.

14. The following data constitute a sample of the time (in minutes) that it takes firemen to reach reported fires. This sample is taken for two different fire stations and includes only one-alarm fires (that station was the only one required at the fire).

Station 1	Station 2
4.5	1.0
5.0	4.5
6.0	9.5
5.5	1.0
3.5	1.5
4.0	9.0
6.0	2.0
5.5	3.0
5.0	1.5
5.0	3.5
	8.0
	3.5

a. Find the mean and the standard deviation of the response time for Station 1.

b. Find the mean and the standard deviation of the response time for Station 2.

c. Assuming that the crews from both fire stations have equal ability once they arrive at the scene, which station area has better fire protection? Why?

15. Without referring to the text, write out your intuitive definition of dispersion of data. Then compare your definition with that given in the text.

16. Suppose you scored in the 55th percentile in a standardized English exam. What would this mean? What would it mean if you were above the 3rd quartile in your class?

SELECTED REFERENCES

Bryant, Edward C. *Statistical Analysis.* Revised edition. New York: McGraw-Hill, 1966.

Guenther, William C. *Concepts of Statistical Inference.* Second edition. New York: McGraw-Hill, 1973. Especially Chapter 3.

Miller, Irwin and John E. Freund. *Probability and Statistics for Engineers.* Englewood Cliffs, N.J.: Prentice-Hall, 1965. Especially Chapter 6.

Spurr, William A. and Charles P. Bonini. *Statistical Analysis for Business Decisions.* Revised edition. Homewood, Ill.: Irwin, 1973. Especially Chapter 4.

Tanur, Judith M., Frederick Mosteller, William H. Kruskal, Richard F. Link, Richard S. Pieters, and Gerald R. Rising. *Statistics: A Guide to the Unknown.* San Francisco: Holden-Day, 1972.

Wallis, W. A. and H. V. Roberts. *The Nature of Statistics.* New York: Free Press, 1962. Especially Chapters 5, 7.

Yamane, Taro. *Statistics: An Introductory Analysis.* Third edition. New York: Harper & Row, 1973. Especially Chapter 4.

APPENDIX
USING CODED DATA TO CALCULATE A STANDARD DEVIATION

FOR THOSE WHO must do their calculations by hand, this appendix presents a shortcut method for finding the standard deviation of large, cumbersome numbers. The shortcut involves coding the original data by subtracting a constant from each observed value or dividing each observed value by a constant, or both. If the original data are coded by subtracting a constant from each observation, the standard deviation does not change. In other words, the two sets of data

1007		7
1005		5
1004	and	4
1006		6
1003		3

both have the standard deviation of 1.58, even though there is a constant difference of 1000 between the sets. Hence the data can be coded by subtracting a constant from each observation, and the standard deviation of the resulting coded data is the standard deviation of the original data.

We can also code the data by dividing each observation by a constant K. Suppose we have the following data:

x
900
5000
3000
1000
100

The standard deviation of these data is 1988.7. If we code the data by dividing each observation by 1000,

x/K
.9
5.0
3.0
1.0
.1

the standard deviation of the coded data is 1.9887. To uncode this we multiply the coded standard deviation (SD) by 1000:

$$SD = 1000 \cdot \text{coded SD}$$
$$= 1000 \cdot 1.9887$$
$$= 1988.7$$

which is the same answer we received previously.

In general, then, if the data are coded by dividing each observation by a constant K, the standard deviation of the original data is found by

$$SD = K \cdot \text{coded SD}$$

If the data are coded by subtracting a constant from each observation, the standard deviation remains unchanged.

PART II
HOW TO COPE WITH UNCERTAINTY

CHAPTER 5
THE BASICS OF PROBABILITY

OBJECTIVES: After studying this chapter and working the problems, you should be able to:

Illustrate the three sources of probability and explain the meaning of each.

Recognize and calculate joint probabilities.

Recognize and calculate conditional probabilities.

Define and give an example of independent events, mutually exclusive events, and collectively exhaustive events.

Demonstrate your ability to use the counting rules for determining how many ways events can happen.

HAVE YOU EVER considered the number of events that will happen with absolute and complete certainty? When you think about this question and look at the world around you, you have to conclude that the number of such events is minimal to say the least. Most people intuitively recognize this lack of certainty and subconsciously sprinkle their conversation with such phrases as "I think so," "probably," "maybe," "I doubt it," and "God willing and the creek don't rise." As the old saying goes, nothing's certain but death and taxes. But with organ transplants even the medical profession is having a hard time determining when someone's dead—and we certainly know that some people avoid paying taxes.

In other words, we must formally recognize something that we subconsciously realize: the world operates on uncertainty rather than certainty. Once we are willing to admit this, we need some means of describing and analyzing this uncertain world. If the world operates on uncertainty, then it seems only logical that *probability* should be used as a tool to describe and analyze the uncertainty that surrounds us.

FUNDAMENTAL CONCEPTS OF PROBABILITY

Probability is basically a measure of relative frequency or a measure of what proportion of the time we expect something to happen. When someone says there's a 50-50 chance that a product will be profitable, he is really saying this: "If we were to introduce many products like this in a market environment exactly like the one we have now, half (50 percent) of them would be profitable and half (50 percent) of them would not be profitable."

Sometimes an intuitive idea of probability or chance doesn't help us. For instance, the statement that "birth control pills are 97 percent effective" may have different meanings. It could mean that in a large group of women relying on the pill, 3 percent of them could be expected to become pregnant. It could also mean that if a woman relied on the pill a large number of times, she could expect to become pregnant about 3 percent of the times. And, finally, it could mean that every time a woman relies on the pill she gets about 3 percent pregnant! Obviously the first interpretation is the one intended. Nevertheless the example points out that we are often exposed to numerical data and make inferences without reflecting on what the data really mean.

ASSIGNING PROBABILITY VALUES

Let's begin formalizing some of our intuitive ideas about probability so that we can use it more effectively. First of all, the probability of an event is *never* less than zero or more than 1. If it is absolutely impossible that an event can occur, then we assign to the probability of that event a value of zero. For instance, the probability of your turning this printed page into a genuine sheet of 14 carat gold is zero. We would write this:

$$\text{Pr(turning this printed sheet into 14 carat gold)} = 0$$

We will use "Pr" to mean "the probability of." The part of the statement within parentheses—turning this printed sheet into 14 carat gold—is the *event*. An event is an occurrence or an outcome of an experiment. Hence if an impossible event has a probability of zero, there is no meaning to any probability less than zero or negative. There is no such thing as a "negative" probability value.

But if the event is guaranteed to happen, we assign it a probability value of 1. If a person has a drawer full of brown socks and black socks, for instance, the probability of getting a matching pair in a randomly drawn group of three socks is 1, or certainty. (Think about that a minute.) We would write this:

$$Pr(\text{matching pair in a random group of 3 socks}) = 1$$

A probability of 1 means the event *has* to happen. There can be nothing greater than that. There is no such thing as a probability value greater than 1.

Let's look at an example with which we are all familiar—the tossing of a fair coin. Ignoring the negligible possibility of landing on edge, we generally accept the idea that a fair coin will land "heads" 50 percent of the time and "tails" 50 percent of the time. This relative frequency of the occurrence of a head is the probability of obtaining a head in one toss of a fair coin. In probability notation we would write

$$Pr(\text{head}) = \tfrac{1}{2}$$

Which is no shocking news to anyone.

To formalize this we consider: if all the things that can happen (events) are equally likely to occur, then we can find the probability of a "success" by using the following formula:

SO&B 5-1: $\quad Pr(\text{success}) = \dfrac{\text{number of ways a success can happen}}{\text{total number of events that can happen}}$

Remember: All events must be equally likely for this formula to work.

In the tossing of a coin, if we want a head to appear then the appearance of a head is a success. The number of ways we can succeed is one, a head. The total number of events that can occur is two, a head or a tail. Using SO&B 5-1,

$$Pr(\text{head}) = \dfrac{\text{number of ways a head can happen}}{\text{total number of events that can happen}} = \tfrac{1}{2}$$

The translation of a probability such as Pr(head) is that in a large number of experiments (tossing of a fair coin) we would expect about one-half of them to result in the appearance of a head. It does *not* mean that we will get exactly one-half of the tosses to be heads. Nor does it mean, of course, that if you toss the coin once you would expect to get one-half of a head!

SOURCES OF PROBABILITY VALUES

One question naturally arises at this point: How do we initially find these relative frequencies or probability values for simple events? We will be looking shortly at how to manipulate probability values, but first we must find a starting point. Probability values can come from any of three different sources. The tossing of a fair coin is *a priori* or mathematical probability. A priori means that we know beforehand, without any experimentation, the value of the probability. We can calculate a priori probabilities purely mathematically. In tossing a fair coin we know that Pr(head) $= \frac{1}{2}$ without tossing the coin 1 million times and counting the number of heads that occurred. In a deck of cards we know various probabilities beforehand without extensive experimentation. We know that Pr(a heart in one draw) $= \frac{1}{4}$ or $\frac{13}{52}$ and that Pr(ace of spades in one draw) $= \frac{1}{52}$ a priori without experimentation. Probabilities involving roulette wheels and raffle tickets are also examples of a priori probabilities. Some students claim that statistics is a priori because they know the probability of their ever understanding statistics is zero and this knowledge comes without any experimentation. I don't really believe that.

When we refer to mortality rates, though, we must gather a large amount of historical data before we can calculate probabilities. This source of probability values is called *empirical*. Empirical probabilities are based on experimentation or large files of historical data. Mortality rates used in the insurance industry are a classic example of empirical probabilities. Based on large numbers of historical death rates, the probability of death occurring at a particular age can be estimated empirically. The probability of defective items in a production process is also found empirically. In a large production run the number of defective items can be tabulated and, by applying SO&B 5-1, the probability of a defective unit occurring can be determined.

The third source of probability values is *subjective*. A subjective probability is based on a person's opinion or belief about a situation. For instance, the decision maker may be faced with a choice of expanding or not expanding an operation. The major variable could be the general economy of the nation. If the economy is going to expand, then he would want to expand his operation. After considering the situation the decision maker may decide there is a 70 percent chance that the nation's economy will expand. This would be a subjective probability because it is not based on precise historical data (even though his past experience with such matters would influence his decision) nor is it based on any mathematical or a priori probability. It is his opinion or belief after looking at the facts and considering past experiences.

Probabilities from subjective sources can be very useful in business, but they are scorned by many mathematicians. The value of a subjective probability is that even though it is subjective, it's often the best you've got. Moreover, the alternative to utilizing subjective probabilities is usually making a definite statement like "The economy *will* expand"—and this is obviously fraught with danger because it tends to make one ignore the fact that there is a chance it won't expand.

The three sources of probability values are thus a priori, empirical, and subjective. All probability values originate in one of these sources.

MANIPULATING PROBABILITIES

Although most people have a basic intuitive idea of probability, these ideas seem to break down when probabilities have to be manipulated by combining or expanding them in some way. One concept that can get in the way of your intuitive idea of probability is that of *independence*. We will formalize the concept of statistical independence later, but right now we can say that two events are independent if they have absolutely no effect on each other. For instance, I just now flipped a coin and a head came up. If you flip a coin right now while you are reading this, the fact that I got a head will have absolutely no effect on how your coin comes up. The event "heads on my coin" and the event "heads on your coin" are statistically independent.

JOINT PROBABILITY

A *joint probability* is the probability of two events occurring. The probability of my getting a head on my coin *and* your getting a head on your coin would be a joint probability. To find the joint probability of two independent events A and B, we use SO&B 5-2:

SO&B 5-2: $\Pr(A \text{ and } B) = \Pr(AB)$

$$= \Pr(A) \cdot \Pr(B)$$

for *independent* events A and B

We will use the notation $\Pr(AB)$ to denote the joint probability of A and B. Let A be the event that I get a head on my coin and let B be the event that you get a head on your coin. Then the joint probability of A and B would be

$$\Pr(AB) = \Pr(A) \cdot \Pr(B)$$

$$= \tfrac{1}{2} \cdot \tfrac{1}{2}$$

$$= \tfrac{1}{4}$$

We could illustrate this also by listing all the possible outcomes. Such a listing or description is called a *sample space*. For this example the sample space would be

Event	My Coin	Your Coin
1	Head	Head
2	Head	Tail
3	Tail	Head
4	Tail	Tail

Each line represents a possible outcome of tossing the two coins. There are four possible outcomes, each of which is equally likely, and one of the outcomes is a success (head and head). Hence, using SO&B 5-1,

$$\Pr(AB) = \tfrac{1}{4}$$

Suppose we are assembling an electronic device with two transistors. The device will work if only one transistor is defective, but it will not work if both transistors are defective. It is known that 10 percent of the transistors being used are defective. The assembler randomly draws the two transistors from a large bin. What proportion of completed devices would we expect not to work? This would be the same as asking, "What is the probability of transistor 1 being defective *and* transistor 2 being defective?" The two events—"the first transistor being defective" and "the second transistor being defective"—are independent events because we are assuming that a large number of transistors are available. Using SO&B 5-2 we could find:

Expected proportion of nonworking devices

$= \mathrm{Pr}$(transistor 1 defective *and* transistor 2 defective)

$= \mathrm{Pr}$(transistor 1 defective) $\cdot \mathrm{Pr}$(transistor 2 defective)

$= .10 \cdot .10$

$= .01$

If we were to look at the probability of selecting at random a transistor that is not defective, this probability would be the *complement* of the probability of a defective item. We know that if we draw a transistor it must be either defective or nondefective. The probability that the selected transistor is defective or nondefective is 1:

Pr(transistor is defective *or* nondefective) $= 1$

Therefore if we want to know the complement of the probability of an event A, we calculate:

SO&B 5-3 $\mathrm{Pr}(\text{not } A) = \mathrm{Pr}(\sim A) = 1 - \mathrm{Pr}(A)$

We denote "not A" by "$\sim$A." Using SO&B 5-3 the probability of a nondefective item is

Pr(nondefective) $= \mathrm{Pr}(\sim \text{defective})$

$= 1 - \mathrm{Pr}$(defective)

$= 1 - .10$

$= .90$

From this point, then, we can calculate the remaining probabilities for the assembly of the electronic devices using SO&B 5-2 and SO&B 5-3:

Pr(transistor 1 defective and transistor 2 defective)

$= .10 \cdot .10$

$= .01$

Pr(transistor 1 nondefective and transistor 2 defective)

$= .90 \cdot .10$

$= .09$

Pr(transistor 1 defective and transistor 2 nondefective)
$= .10 \cdot .90$

$= .09$

Pr(transistor 1 nondefective and transistor 2 nondefective)
$= .90 \cdot .90$

$= .81$

CONDITIONAL PROBABILITY

We can now introduce the *conditional probability*. Conditional probability deals with finding the probability of an event under the condi-
tion that another event has already occurred. This is written $Pr(A|B)$,
which means the probability of event A given that event B has already
occurred.

Suppose in our transistor example that instead of having a large
bin full of transistors, we have only three transistors from which to
choose and we know that exactly *one* of the three is defective. If one
of the three is defective, then when the assembler selects his first
transistor, the probability of that transistor being defective is (using
SO&B 5-1)

$$Pr(\text{1st transistor defective}) = \tfrac{1}{3}$$

But suppose that after selecting that first transistor the assembler
checks it and finds that it is *not* defective. When he draws the second
transistor he knows there is still one defective left and only two
transistors from which to choose. Therefore the probability of the
second transistor being defective under the condition that the first
was not defective is (using SO&B 5-1)

$$Pr(\text{2nd transistor defective}|\text{1st transistor nondefective}) = \tfrac{1}{2}$$

Let's put this in a formal statement we can use when the answers
cannot be observed. The formula for the conditional probability of
event A given that event B has happened is

SO&B 5-4: $$Pr(A|B) = \frac{Pr(AB)}{Pr(B)}$$

For this example let A be the event that the second transistor selected
if defective and let B be the event that the first transistor selected is
nondefective. Then, using SO&B 5-4, we want to find

$$\Pr(A|B) = \frac{\Pr(AB)}{\Pr(B)}$$

or

Pr(2nd transistor defective|1st transistor nondefective)

$$= \frac{\text{Pr(2nd transistor defective } and \text{ 1st transistor nondefective)}}{\text{Pr(1st transistor nondefective)}}$$

There are three distinct but equally likely events that can occur in drawing the two transistors. This is shown in the following sample space:

Possible Event	1st Transistor Is:	2nd Transistor Is:	Remaining Transistor Is:	
1	Nondefective	Defective	Nondefective	←AB
2	Nondefective	Nondefective	Defective	←B
3	Defective	Nondefective	Nondefective	

We can see that only one of three possible events satisfies both A and B. Therefore, from SO&B 5-1,

$$\Pr(AB) = \tfrac{1}{3}$$

We can also see that two of the three events satisfy event B. Hence

$$\Pr(B) = \tfrac{2}{3}$$

Using SO&B 5-4 we would then calculate

$$\Pr(A|B) = \frac{\Pr(AB)}{\Pr(B)}$$

$$= \frac{\tfrac{1}{3}}{\tfrac{2}{3}}$$

$$= \tfrac{1}{2}$$

which agrees with our prior observed value.

THE RULE OF ADDITION

We need to look at one more rule for manipulating probabilities — the *rule of addition*. The rule of addition concerns the probability of A or B, where A and B are two events. First we need to define the term *or* as used in statistics and other disciplines. The probability of A or B, Pr(A or B), is the probability of A happening or B happening or both A and B happening. The rule of addition is

SO&B 5-5: $\Pr(A \text{ or } B) = \Pr(A) + \Pr(B) - \Pr(AB)$

We can demonstrate this rule with our example of transistors. Again we have a large bin of transistors, 10 percent of which are defective. We are assembling them into units that require two transistors. Let A be the event that the first transistor is defective and let B be the event that the second transistor is defective. Now we would like to know the probability that the assembled unit contains a defective transistor. This would be the probability that the first is defective but not the second, or the second is defective but not the first, or both are defective. In other words, we are looking for Pr(A or B). We know from prior work that

$$\Pr(A) = .10$$
$$\Pr(B) = .10$$
$$\Pr(AB) = .10 \cdot .10 = .01$$

Therefore, using SO&B 5-5, we have

$$\Pr(A \text{ or } B) = \Pr(A) + \Pr(B) - \Pr(AB)$$
$$= .10 + .10 - .01$$
$$= .19$$

Of all the units assembled containing two transistors, we would expect 19 percent of them to have a defective transistor.

VENN DIAGRAMS

The additive rule sometimes leaves a student wondering how it came about. We can demonstrate the additive rule through a *Venn diagram* — a schematic way of showing probability. Figure 5-1 shows a Venn

FIGURE 5-1. Probability of
Defective Transistors in
Assembly Containing Two
Transistors

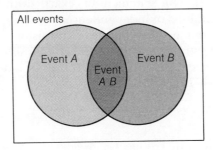

diagram concerning our problem with the transistors. If we want to
know $\Pr(A$ or $B)$ and we add $\Pr(A) + \Pr(B)$, then we will have added
the cross-hatched area, $\Pr(AB)$, twice. But we only want it once, so
we subtract it out once to arrive at $\Pr(A$ or $B) = \Pr(A) + \Pr(B) - \Pr(AB)$.

In our example both event A and event B can happen. Therefore,
in the Venn diagram these events are shown to overlap as $\Pr(AB)$.
If the two events A and B could not both happen, we call them *mutually exclusive* events and in our Venn diagram they will appear as
nonoverlapping circles. If we let C be the event that the first transistor is defective and let D be the event that the first transistor is nondefective, then event C and event D are mutually exclusive since,
obviously, if the first transistor is defective it cannot be nondefective
and vice versa.

We would also say that events C and D are *collectively exhaustive*
because they represent *everything* that can happen with the first
transistor. The sum of the probabilities of a group of mutually exclusive and collectively exhaustive events equals 1. Notice that if we
do not know whether or not the first transistor is defective, then the
probability that the second transistor is defective is still $\frac{1}{3}$. Knowledge
of the result of the first draw, though, changes our probability concerning the second.

DETERMINING STATISTICAL INDEPENDENCE

At this point we can formalize our concept and definition of *independence*. We now formally define statistical independence as follows:
Two events, A and B, are independent if and only if

$$\Pr(A) = \Pr(A|B)$$

We can see that this equation is merely telling us that the probability of event A remains the same regardless of whether or not event B has happened. This statement fits exactly with our previous intuitive, informal definition of independence.

In the example with only three transistors, one of which is defective, let B be the event that the first transistor is nondefective and let A be the event that the second transistor is defective. Then without knowledge of B,

$$\Pr(A) = \tfrac{1}{3}$$

But if we know that B has occurred,

$$\Pr(A|B) = \tfrac{1}{2}$$

Since $\Pr(A) \neq \Pr(A|B)$ because $\tfrac{1}{3} \neq \tfrac{1}{2}$, events A and B are *not* independent or are *dependent*. This formalized definition, including its determining equation, is necessary because the analyst cannot usually tell by mere intuition whether or not two events in real-life situations are independent. But by using the formal definition the analyst can calculate $\Pr(A)$ and $\Pr(A|B)$, compare the two to see whether they are equal, and thus know whether the two events are independent or not.

With the tools for manipulating probabilities that we now have, we can generalize our formula for calculating a joint probability. Recall that SO&B 5-2 is valid only when event A and event B are independent. We had SO&B 5-2:

$$\Pr(AB) = \Pr(A) \cdot \Pr(B) \qquad \text{for independent events } A \text{ and } B$$

We now generalize that formula to SO&B 5-6:

SO&B 5-6:
$$\Pr(AB) = \Pr(A) \cdot \Pr(B|A)$$
$$= \Pr(B) \cdot \Pr(A|B)$$
$$= \Pr(BA)$$

Let's return to our example of the three transistors with one defective. Knowing that

$$\Pr(\text{1st transistor nondefective}) = \tfrac{2}{3}$$

and

Pr(2nd transistor defective|1st transistor nondefective) = $\frac{1}{2}$

then, using SO&B 5-6,

> Pr(1st transistor nondefective *and* 2nd transistor defective)
> = Pr(*AB*)
> = Pr(*B*) · Pr(*A*|*B*)
> = $\frac{2}{3} \cdot \frac{1}{2}$
> = $\frac{1}{3}$

which agrees with the answer we previously observed.

Now let's digress for a moment and consider the plight of Weird Willy the statistician. Willy does a lot of traveling by airplane doing consulting work and speaking to professional organizations, but on one of his trips he read a research article that claimed there was one chance in a thousand of a bomb being on any particular airplane flight. Willy did some quick figuring and, sure enough, he had made about a thousand plane trips. Suddenly he decided not to travel by plane any more. Willy would spend days traveling across the country by bus rather than take any more chances on the airplane. Then he read another article in a technical journal that showed there was only one chance in a *million* that there would be two bombs on the same plane. Weird Willy immediately went back to flying . . . but now he always carries a bomb in his briefcase.

This anecdote demonstrates a few principles. First of all, the fact that Willy had made about a thousand trips should not have been compared to the "one in a thousand" chance because each flight is *independent* of the others. Every time he got on a plane the chance would still be one in a thousand regardless of what had happened on previous flights. Secondly, Willy seems to have gotten confused about joint probability and conditional probability. The "one in a million" chance of two bombs being on a plane is a *joint* probability. Once it is known that one bomb is on the plane, we have a conditional probability. Now let's look at another example of probability in action.

> **Example 1:** An old acquaintance of ours, Mabel Metcher, is considering buying an automobile, and her friend Finley runs a used car lot. Finley is currently conducting a

special promotion: for $2000, the car buyer signs a contract to purchase whatever car he draws from a large basket. Every car in the lot is represented in the basket by a ticket bearing the serial number. Finley knows that some of the cars have no major defects. If a car does have a major defect, it is in the engine or in the transmission (or both).

Mabel asks Finley if he will just tell her what cars are good instead of having to draw a number from the basket. Finley refuses, but he does agree to tell Mabel how many cars he has with no defects, defective engines, or defective transmissions—and, in addition, how many of the 50 cars he has on the lot are less than five years old or are five years or older. He prepares for her the data shown in Table 5-1.

TABLE 5-1.   Classification of Cars on Finley's Lot

	Total	Defective Engine	Defective Transmission	No Major Defects
Less than 5 years old	20	11	8	9
5 years or older	30	14	12	6
	50	25	20	15

Mabel knows that any car which is less than five years old and has no defects is worth more than the $2000 it would cost her. She could then calculate the probability of drawing such a car:

$$\text{Pr(less than 5 years old } and \text{ no defects)} = \tfrac{9}{50}$$

$$= .18$$

Mabel also wants to know the odds of getting a car that has both a defective engine and a defective transmission. Looking at the bottom row of totals we see that of the total of 50 cars, 15 have no major defects. Therefore 50 minus 15 leaves 35 cars with a major defect. But Finley has shown that there are 25 cars with a defective engine and 20 cars with a defective transmission. We can calculate, then, that

      25 defective engines
  +20 defective transmissions
      45 total defects
  −35 cars with a major defect
      10 cars with both a defective engine *and* a defective
         transmission

What we have accomplished here is to break the classification of defects into collectively exhaustive and mutually
exclusive events as shown in Table 5-2.

**TABLE 5-2. Classification of Cars by
Type of Defect**

Type of Defect	No. of Cars
No major defect	15
Defective engine only	15
Defective transmission only	10
Defective engine and transmission	10
	50

Mabel can now see that the probability of drawing a car
with both a defective engine and a defective transmission is

$$\text{Pr(defective engine } and \text{ defective transmission)} = \tfrac{10}{50}$$

$$= .20$$

   For an additional $500, Finley will allow Mabel to draw a
number from just those cars that are less than five years
old. Once again Mabel wants to know her chances of
drawing a car with no defects. Now she would be looking
at a conditional probability:

$$\text{Pr(no major defects} | \text{less than 5 years old)}$$

We have already found that Pr(less than 5 years old *and*
no major defects) = .18. By looking back at Table 5-1 we
see that

$$\text{Pr(less than 5 years old)} = \tfrac{20}{50}$$

$$= .40$$

Therefore

Pr(no major defects|less than 5 years old)

$$= \frac{\text{Pr(no major defects and less than 5 years old)}}{\text{Pr(less than 5 years old)}}$$

$$= \tfrac{.18}{.40}$$

$$= .45$$

Are the events "no major defects" and "less than five years old" independent? Looking at Table 5-1 we see that

$$\text{Pr(no major defects)} = \tfrac{15}{50}$$

$$= .30$$

We just calculated, though, that

$$\text{Pr(no major defects|less than 5 years old)} = .45$$

Then $\text{Pr}(A) \neq \text{Pr}(A|B)$ since $.30 \neq .45$, and the two events are not independent.

Mabel may also be interested in knowing the probability of getting a car with a defect. This would be Pr(defective engine *or* defective transmission), which we could find by

Pr(defective engine *or* defective transmission)

$= \text{Pr(defective engine)} + \text{Pr(defective transmission)}$

$\quad - \text{Pr(defective engine and defective transmission)}$

$$= \tfrac{25}{50} + \tfrac{20}{50} - \tfrac{10}{50}$$

$$= \tfrac{35}{50}$$

$$= .70$$

In this case we could have found the answer in a different way. We know that all the cars either have a defect or do not. We already know that Pr(no major defect) $= .30$. The complement of this would be Pr(major defect). Therefore we could have calculated

$$Pr(\text{major defect}) = 1 - Pr(\text{no major defect})$$
$$= 1 - .30$$
$$= .70$$

which agrees with our first calculation.

DETERMINING THE NUMBER OF EVENTS

We have seen that if all events are equally likely, we can find the probability of a success by forming the ratio

$$Pr(\text{success}) = \frac{\text{number of ways a success can happen}}{\text{total number of events that can happen}}$$

This is all well and good except for one thing: sometimes it's tedious to count the number of events that can happen. We will therefore look at a few shortcuts in "counting how many."

Suppose we had to staff a new branch office with an office manager, a secretary, and an order clerk and we had four candidates for the office manager's position, two candidates for secretary, and three candidates for order clerk. We could find the number of ways to staff the branch office by using the *multiplicative rule*. The multiplicative rule states that if we have a equally likely choices for event A, b equally likely choices for event B, c equally likely choices for event C, d equally likely choices for event D, and so forth, then the total number of separate combinations that can be made from these events is

SO&B 5-7:   Total combinations $= a \cdot b \cdot c \cdot d \cdots$

Using the multiplicative rule (SO&B 5-7), we could calculate that the total number of ways to staff the branch office is

$$\text{Total combinations} = a \cdot b \cdot c$$
$$= 4 \cdot 2 \cdot 3$$
$$= 24$$

As an exercise you might try listing all possible combinations to verify that there are, in fact, 24 of them.

If three friends had applied for jobs at the branch office—Carlos as office manager, Rosie as secretary, and John as order clerk—and all applicants were equally likely to be chosen, the probability of all three of them being hired to staff the new office would be (using SO&B 5-1)

$$\text{Pr(all three friends being hired)} = \tfrac{1}{24}$$

since there is only one combination that includes all three friends and there are 24 total combinations.

In using the multiplicative rule we had four people applying for the first job, two people applying for a distinctly different second job, and three people applying for a distinctly different third job. Suppose, though, that the new branch office was on a tropical isle in the South Pacific and all nine people were applying for and would accept any of the three jobs just to get moved to the new location. Now we are presented with a different situation in "counting the ways" we can staff the new branch office with an office manager, a secretary, and an order clerk. To simplify our counting procedure here we would use the *permutation rule.* The permutation rule states that if we are choosing *r* items from a group of *n* items and the *order* in which the *r* items are chosen is important, then the total number of ways this can be chosen is found by

SO&B 5-8 $$_nP_r = \frac{n!}{(n-r)!}$$

The symbol $_nP_r$ is called the number of *permutations* of *n* things taken *r* at a time. The ! is read "factorial" and

$$n! = 1 \cdot 2 \cdot 3 \cdots (n-1) \cdot n$$

or

$$0! = 1$$

$$1! = 1$$

$$2! = 1 \cdot 2 = 2$$

$$3! = 1 \cdot 2 \cdot 3 = 6$$

$$4! = 1 \cdot 2 \cdot 3 \cdot 4 = 24$$
$$5! = 1 \cdot 2 \cdot 3 \cdot 4 \cdot 5 = 120$$

and so forth.

In staffing our new tropical office, we are choosing three people from a group of nine and the order of choice is important. In other words, the staff

> Office manager: Carlos
>
> Secretary: Rosie
>
> Order clerk: John

is distinctly different from the staff

> Office manager: Rosie
>
> Secretary: John
>
> Order clerk: Carlos

We could then find the number of ways to staff the new office to be (using SO&B 5-8):

$$_nP_r = {_9}P_3 = \frac{9!}{(9-3)!}$$

$$= \frac{9!}{6!}$$

$$= \frac{9 \cdot 8 \cdot 7 \cdot 6 \cdot 5 \cdot 4 \cdot 3 \cdot 2 \cdot 1}{6 \cdot 5 \cdot 4 \cdot 3 \cdot 2 \cdot 1}$$

$$= 9 \cdot 8 \cdot 7 \quad \text{since the} \\ (6 \cdot 5 \cdot 4 \cdot 3 \cdot 2 \cdot 1)\text{'s} \\ \text{cancel each other}$$

$$= 504$$

How many ways could Carlos, Rosie, and John be sent to staff the new office? Here we would be finding the number of ways we can staff the three positions from a group of three (with order still important). So, using SO&B 5-8,

$$_nP_r = {}_3P_3 = \frac{3!}{(3-3)!}$$

$$= \frac{3!}{0!}$$

$$= \frac{3 \cdot 2 \cdot 1}{1}$$

$$= 6$$

Hence if all choices are equally likely, the probability that the three-person staff will be Carlos, Rosie, and John is

$$\text{Pr(staff is Carlos, Rosie, John in any positions)} = \tfrac{6}{504}$$

$$= .012$$

Now we are ready to examine the third shortcut in counting. Let's just expand the example one step further by eliminating the three distinct positions of office manager, secretary and order clerk. Now the office will be staffed with three people—all on an equal status. The staff Carlos, Rosie, John is now identical to the staff Rosie, John, Carlos. In other words, the order of selection is no longer of any importance. To shortcut our counting procedure here we would use the *combination rule*. The combination rule states that if we are choosing r items from a group of n items and the order is *not* important, then the total number of ways this can be chosen is found by

SO&B 5-9: $$_nC_r = \frac{n!}{r!(n-r)!}$$

The symbol $_nC_r$ is called the number of *combinations* of n things taken r at a time.

In our new staffing problem, we are staffing the branch office with three people. They have no titles, so order is not important. Hence the number of ways we can make up our new staff from the nine candidates is (using SO&B 5-9):

$$_nC_r = {}_9C_3 = \frac{9!}{3!\,6!}$$

$$= \frac{9 \cdot 8 \cdot 7 \cdot 6 \cdot 5 \cdot 4 \cdot 3 \cdot 2 \cdot 1}{(3 \cdot 2 \cdot 1) \cdot (6 \cdot 5 \cdot 4 \cdot 3 \cdot 2 \cdot 1)}$$

$$= \frac{9 \cdot 8 \cdot 7}{3 \cdot 2 \cdot 1} \qquad \text{since the } (6 \cdot 5 \cdot 4 \cdot 3 \cdot 2 \cdot 1)\text{'s}$$
$$\text{cancel each other}$$

$$= 3 \cdot 4 \cdot 7$$

$$= 84$$

The number of ways Carlos, Rosie, and John could be sent to staff the office would be (again using SO&B 5-9):

$$_nC_r = {_3C_3} = \frac{3!}{3!(3-3)!}$$

$$= \frac{3!}{3!(0!)}$$

$$= \frac{3 \cdot 2 \cdot 1}{(3 \cdot 2 \cdot 1) \cdot 1}$$

$$= 1$$

Hence the probability of the three friends being sent together to staff the office, if all candidates are equally likely to be selected, is

$$\text{Pr(staff is Carlos, Rosie, John)} = \tfrac{1}{84}$$

$$= .012$$

AUTOPSY

The first introduction to probability is often confusing to the beginning statistical analyst. You could spend the rest of your life studying probability. Although it is not my purpose in this chapter (or the next) to make you an expert in probability, you should be developing a basic idea of the fundamentals of probability and how to use them. You should feel successful at this point if you have become more sensitive to the uncertainties of your environment. The "how to" part of the fundamentals will come to you as you work with them. The problems at the end of this chapter are designed to give you some practice.

SUMMARY

Important Terms

Probability ranges in value from 0 to 1.

If $\Pr(A) = 1$, then A will certainly happen.

If $\Pr(A) = 0$, then A cannot happen.

The three sources of probability are a priori, empirical, subjective.

If all events are equally likely, then

$$\Pr(\text{success}) = \frac{\text{number of ways}}{\text{total number of events}}$$

$\Pr(\text{not } A) = 1 - \Pr(A)$.

The joint probability of two events A and B is

$$\Pr(AB) = \Pr(A)\,\Pr(B\,|\,A) = \Pr(B)\,\Pr(A\,|\,B) = \Pr(BA)$$

Remember: Joint probability usually requires multiplication of probabilities.

The probability of event A or event B happening is

$$\Pr(A \text{ or } B) = \Pr(A) + \Pr(B) - \Pr(AB)$$

Remember: An *or* probability usually requires adding and subtracting probabilities.

The probability of event A under the condition that event B has happened is

$$\Pr(A\,|\,B) = \frac{\Pr(AB)}{\Pr(B)}$$

Independence: Two events A and B are independent if and only if $\Pr(A) = \Pr(A\,|\,B)$ or $\Pr(B) = \Pr(B\,|\,A)$.

Mutually exclusive: Events are mutually exclusive if the occurrence of one precludes the occurrence of the other.

Collectively exhaustive: A group of events is collectively exhaustive if they include *all* events that can possibly happen.

Counting Rules

Number of ways to select one each from a number of groups: $a \cdot b \cdot c \cdot d \cdots$.

Number of ways to select r items from a group of n (order important): $_nP_r$

Number of ways to select r items from a group of n (order not important): $_nC_r$

PROBLEMS

1. In discussing the fate of next year's budget the president of a state college made the following observation: "I think there is a 60–40 chance the legislature will approve our budget based on conversations with key representatives. However, analysis of the prior records shows that the legislature has approved 65 of the last 100 similar requests. I have heard that the governor may take the 50 requests that are undecided yet, such as ours, and randomly draw 25 to be approved." Identify the a priori, the empirical, and the subjective probabilities in this statement. Explain your choices.

2. A camera distributor has 100 model I's. He remembers that he scavenged an internal part from eight of them to handle warranty complaints. In randomly drawing a camera of this model, what is the probability that it will be defective (one that has been scavenged)? What is the probability that it will not be one of those that have been scavenged?

3. Of 1000 accounts receivable in a firm, 600 are current, 250 are 30 days past due, 100 are 60 days past due, and 50 are 90 days or more past due. An auditor draws one statement from accounts receivable. What is the probability that it is current? Thirty days past due? Sixty days past due? Ninety days or more past due? Are the events current, 30 days past due, 60 days past due, and 90 days or more past due mutually exclusive? Collectively exhaustive?

4. From Problem 3, past experience has shown that 5 percent of the current receivables, 10 percent of the 30 days past due, 25 percent of the 60 days past due, and 50 percent of the 90 days or more past due will have to be written off as bad debts.

> *a.* What is the probability that the auditor will draw an account that will end up being a bad debt?
>
> *b.* What is the probability that the auditor will draw an account that is 60 days past due and will be a bad debt?
>
> *c.* What is the probability that the auditor will draw an account that is 60 days past due or will be a bad debt?

d. The auditor pulls one account and recognizes it as a person who *never* pays his bills. What is the probability that it is 60 days past due?

5. You are a new sales manager and have ten salesmen working for you. You have not had time to get acquainted with all your salesmen, but you do know the following facts from the personnel file:

Four always tell dirty jokes to their client.

Six always offer their client a drink.

Three of the dirty-joke tellers are also ones who always offer their client a drink.

Two don't drink and don't tell dirty jokes, but can't stand selling to women.

You get a request from a client for a salesman to call. The purchasing agent for this client is a woman and she is president of the local Women's Temperance Union. You randomly pick a salesman. What is the probability that the salesman won't offend the purchasing agent?

6. In a large manufacturing company, 15 percent of the fire extinguishers are defective. Because the company is large and has numerous plants, there is a great number of fire extinguishers in the company. A union official periodically checks the extinguishers to see whether they are working. If the union official finds two defective extinguishers, he will call a strike.

a. In checking one extinguisher, what is the probability that it functions?

b. In checking two extinguishers, what is the probability that a strike will be called?

7. A man has the following wardrobe:

4 jackets:	5 shirts:	3 slacks:
1 blue	1 white	1 black
1 red	1 blue	1 brown
1 yellow	1 brown	1 blue
1 brown	1 red	
	1 green	

Suppose he dresses early in the morning when it is still dark and doesn't turn on a light so as not to disturb his wife.

  a. How many different outfits can he put on? (An outfit is a jacket, a shirt, a pair of slacks.)

  b. What is the probability that he'll wear a blue jacket?

  c. What is the probability that he'll wear a blue jacket and a white shirt?

  d. What is the probability that he'll wear a blue jacket, a white shirt, and blue pants?

  e. If he has already selected a blue jacket, what is the probability that he'll wear a white shirt?

  f. Are the events "blue jacket" and "white shirt" independent?

8. There will be a drawing at the annual office Christmas party. First prize is a day off, second prize is a free lunch at the company cafeteria, and third prize is an autographed picture of the boss. There are eight people in the office. How many ways can the prizes be distributed?

9. At the office party in Problem 8, two people's names will be drawn to organize the party and clean up after. How many ways can this committee of two be drawn?

10. A company has decided to form an ad hoc committee to determine ways of increasing morale. One person will be chosen randomly from each of the three departments in the company. There are seven people in the sales department, five people in the service department, and three in the financial department. Some of the employees get together over coffee and decide that all three members of the committee must be sharp if the committee is to be of any significance. They think there are two sharp people in sales, three sharp people in service, and two sharp people in finance. What is the probability of having a significant committee?

11. A discount store is putting up a display of fan belts. There are 15 popular sizes of belts, but the display will hold only 10. How many different ways could they set up the display?

12. An oil company has compiled the following information from sales at its service stations:

	Male Customer	Female Customer
Cash	20%	10%
Credit	40%	30%

a. What is the probability that a given customer will be a male paying cash?

b. What is the probability that a given customer is female?

c. What is the probability that a given customer is paying cash?

d. A female customer arrives at a service station. What is the probability that she will pay cash?

e. Is the sex of the customer independent of whether they pay cash or credit?

13. You are considering buying a franchise for the Spastic Spaghetti Spa. The major item sold, of course, is spaghetti. Records show that 70 percent of the customers order spaghetti. For an additional 25 cents the customer can get a plate and silverware. Sixty percent of the customers who order spaghetti also order a plate and silverware. Of those who do not order spaghetti, 20 percent order a plate and silverware. Now suppose there are 1000 customers.

a. How many orders of spaghetti would you expect to sell?

b. How many orders would you expect for spaghetti *and* a plate and silverware?

c. How many customers would you expect to order a plate and silverware?

14. You are the sales manager of a company that sells home burglar alarm systems door to door. You are just starting your enterprise in a new city. From your past sales experience in other cities, you know that your salesmen can make a sale to 20 percent of the homes. You also know that 15 percent of all homes already have a double-safety locking system and will buy a burglar alarm. A recent survey shows that 60 percent of the homes have a double-safety lock.

a. If your sales force makes 50 calls per day, how many sales would you expect per day?

b. You purchase a list of 1000 addresses that are known to have double-safety locks. How many sales would you expect from this list?

c. Are the events "buy a burglar alarm" and "own a double-safety lock" independent?

d. If each call is independent of all others, what is the probability of a salesman making at least one sale in five random calls? (*Hint:* Use the complement.)

e. A local cat burglar purchases from you the addresses of the homes that did not buy a burglar alarm. However, he cannot pick a double-safety lock. What percentage of your list can be successfully burglarized (assuming he will be successful wherever he does not find the double-safety lock)?

15.   Two swinging bachelors, Jake and Sid, share a pad. Because of their active night life, they frequently rely on Alka-Seltzer in the morning. Sid ran out of Alka-Seltzer and has been stealing some from Jake's jar. Whenever he does this, he replaces the stolen disc with a white poker chip. Right now there are six Alka-Seltzers and four chips in Jake's jar. If Jake should draw a chip on some horrendous morning, he will be left mumbling to himself, "Fizz, darn you, fizz!" If this should happen two mornings in a row, he will probably catch on and beat Sid severely about the head and shoulders.

a. Find the probability that the first tablet drawn is an Alka-Seltzer.

b. Find the probability that the first tablet drawn is a poker chip.

c. If the first tablet drawn is a chip, what is the probability that the second one drawn will be an Alka-Seltzer?

d. If the first tablet drawn is a chip, what is the probability that the second will be a chip?

e. Find the probability that the first *and* the second tablets drawn are both poker chips.

16.   A survey of records reveals that ten people died of a certain disease while they were receiving a new treatment for that disease. Another ten people who were not receiving the new treatment died of the same disease. The conclusion drawn by the examiners is that the chances of the disease being fatal are the same regardless of

whether the patient is receiving the new treatment. Is this conclusion valid?

SELECTED REFERENCES

Bryant, Edward C. *Statistical Analysis.* Revised edition. New York: McGraw-Hill, 1966.

Campbell, Stephen K. *Flaws and Fallacies in Statistical Thinking.* Englewood Cliffs, N.J.: Prentice-Hall, 1974. Especially Chapter 11.

Guenther, William C. *Concepts of Statistical Inference.* Second edition. New York: McGraw-Hill, 1973. Especially Chapter 1.

Miller, Irwin and John E. Freund. *Probability and Statistics for Engineers.* Englewood Cliffs, N.J.: Prentice-Hall, 1965. Especially Chapter 2.

Spurr, William A. and Charles P. Bonini. *Statistical Analysis for Business Decisions.* Revised edition. Homewood, Ill.: Irwin, 1973. Especially Chapter 5.

Tanur, Judith M., Frederick Mosteller, William H. Kruskal, Richard F. Link, Richard S. Pieters, and Gerald R. Rising. *Statistics: A Guide to the Unknown.* San Francisco: Holden-Day, 1972.

Yamane, Taro. *Statistics: An Introductory Analysis.* Third edition. New York: Harper & Row, 1973. Especially Chapter 5.

CHAPTER 6
PROBABILITY DISTRIBUTIONS

OBJECTIVES: After studying this chapter and working the problems you should be able to:

Explain why probability distributions (binomial, hyper-geometric, and normal) are useful.

Describe the conditions under which each of the three distributions can be used for probability calculations.

Explain the difference between continuous and discrete variables.

Calculate probabilities for each type of distribution using the appropriate table.

Define and calculate the standard normal deviate.

Draw a normal distribution curve and use it to illustrate probability calculations.

Calculate binomial probabilities using the normal approximation of the binomial.

IN CHAPTER 5 we looked at some basic concepts and manipulations of probability in general. When a complex problem arises, probability manipulations can become very tedious and taxing. Fortunately there are a number of probability distributions that, once recognized, reduce the complexity to a much simpler calculation. We will look at four probability distributions that commonly arise in business situations. Studying these distributions and learning how to recognize them will help you to cope with uncertainty.

DISTRIBUTIONS WITH DISCRETE VARIABLES

We will look first at three distributions of *discrete* variables. Discrete variables can take on only a limited or countable number of values. If we plot the values of a discrete variable on a graph or chart, it will appear as a series of dots with nothing between each pair of dots. But *continuous* variables can take on any value in an interval. If we plot the possible values of a continuous variable on a graph or chart, it will appear as a solid line because it can take on any value in its interval. The number of defects produced by a process is an example of a discrete variable because it can only take on whole numbers. But the speed (in rpm) of a machine is a continuous variable because it can take on *any* fractional value in a given interval. Later in the chapter we will look at a distribution using continuous variables.

THE BINOMIAL DISTRIBUTION

In business we are sometimes faced with the problem of finding probabilities associated with a situation in which (1) there will be only two possible outcomes in each of a known number of trials of the experiment in which (2) the probability of a success remains constant for all trials and (3) each trial is independent of all other trials. Consider the following two situations.

Situation 1: A salesman is going to call on five new accounts. He knows that 60 percent of the people he calls on have already heard of his product. He may be interested in knowing the chances of zero, one, two, three, four, or all five of these new accounts having previously heard of his product. A would-be analyst sometimes has the initial reaction that if 60 percent of the people have heard of his product and he is calling on five accounts, then obviously the number that have heard of his product is 60 percent of 5, or 3. This is merely an "average" value, though. You can see this intuitively by considering that even though 60 percent of the people in a large group may be men, it certainly would not be unusual to find five women talking together within that group.

In this situation (1) there are only two possible outcomes (either the client has heard of the salesman's product or he has not) in each of the five trials, (2) there is a 60 percent chance that each client has heard of the product, and (3) we can safely assume that each trial is independent of the others so long as the new accounts are not related in some strong fashion. We could find the desired probabilities by applying the binomial distribution.

Situation 2: A retail merchant is selling flashlight batteries. He knows from past experience that 5 percent of this type of battery are defective. A good customer buys four batteries. The merchant may be interested in knowing what percentage of such customers will be dissatisfied by having bought defective merchandise.

In this situation (1) there are only two possible outcomes (defective or not defective) for each of the four batteries, (2) the probability of each battery being defective is 5 percent, and (3) each battery is independent of the other three. We could find the desired percentages (probabilities) by applying the binomial distribution.

Let's formalize our ideas about a binomial distribution now. Assume that the following three characteristics are true:

1. There are only two possible outcomes.

2. The probability of a success remains constant from trial to trial.

3. The trials are independent of each other.

Then the probability of having r successes in n trials can be found by using the binomial distribution. Here is the formula for finding r successes in n trials in a binomial distribution:

SO&B 6-1: $\Pr(r \text{ successes} | n \text{ trials}, p) = {}_nC_r p^r q^{n-r}$

where p is the probability of a success on each trial and q is the probability of a failure ($q = 1 - p$).

In situation 1, we have a salesman making five independent calls ($n = 5$). The probability of a person knowing of his product is 60 percent ($p = .60$) for each of the five calls. The probability of a person not knowing of his product is .40 ($q = 1 - .60 = .40$), and each event (client) is independent of the others. We can then find the probability that zero, one, two, three, four, or five of the new clients have already heard of his product by SO&B 6-1.

The probability that none of the clients has heard of his product is

$$\Pr(r = 0 | n = 5, p = .60) = {}_5C_0 \cdot .60^0 \cdot .40^5$$

$$= \frac{5!}{5!0!} \cdot 1 \cdot .0102$$

$$= .0102$$

The probability that exactly one of the clients has heard of his product is

$$\Pr(r = 1 \,|\, n = 5, \, p = .60) = {}_5C_1 \cdot .60^1 \cdot .40^4$$

$$= \frac{5!}{4!1!} \cdot .60 \cdot .40^4$$

$$= .0768$$

The probability that exactly two of the clients have heard of his product is

$$\Pr(r = 2 \,|\, n = 5, \, p = .60) = {}_5C_2 \cdot .60^2 \cdot .40^3$$

$$= \frac{5!}{2!3!} \cdot .60^2 \cdot .40^3$$

$$= .2304$$

The probability that exactly three of the clients have heard of his product is

$$\Pr(r = 3 \,|\, n = 5, \, p = .60) = {}_5C_3 \cdot .60^3 \cdot .40^2$$

$$= \frac{5!}{3!2!} \cdot .60^3 \cdot .40^2$$

$$= .3456$$

The probability that exactly four of the clients have heard of his product is

$$\Pr(r = 4 \,|\, n = 5, \, p = .60) = {}_5C_4 \cdot .60^4 \cdot .40^1$$

$$= \frac{5!}{4!1!} \cdot .60^4 \cdot .40^1$$

$$= .2592$$

The probability that all five of the clients have heard of his product is

$$\Pr(r = 5 \,|\, n = 5, \, p = .60) = {}_5C_5 \cdot .60^5 \cdot .40^0$$

$$= \frac{5!}{5!0!} \cdot .60^5 \cdot .40^0$$

$$= .0778$$

In situation 2, we have a customer buying four batteries ($n = 4$). The probability of any battery being defective is .05 ($p = .05$) for each of the four batteries. Therefore the probability of a battery being good is .95 ($q = 1 - .05 = .95$), and each of the four trials (batteries) is independent of the other three. We can find the probability that the customer gets at least one defective battery by using SO&B 6-1. Notice that the events zero defective, one defective, two defective, three defective, and four defective are mutually exclusive and collectively exhaustive. "At least one defective" means one, two, three, or four defective. We can find the probability of at least one defective by calculating $1 - \Pr(0 \text{ defective})$. Using SO&B 6-1 for this, we find:

$$\Pr(\text{at least 1 defective} | n = 4, p = .05)$$
$$= 1 - \Pr(r = 0 | n = 4, p = .05)$$
$$= 1 - {}_4C_0 \cdot .05^0 \cdot .95^4$$
$$= 1 - .8145$$
$$= .1855$$

The merchant may decide that if 18.55 percent of the customers buying four batteries will be dissatisfied, it would be too costly in terms of goodwill. Therefore he should stock a more reliable battery.

So long as the three characteristics listed above are satisfied, a binomial distribution can be utilized. There are many situations in the real world that have these three characteristics. The key for you is to watch for these three characteristics so that you can recognize the binomial distribution.

To ease your pain, the probabilities associated with the binomial distribution have been tabulated in Table A-1 at the end of this book. Tables can be found elsewhere that include more values for p and higher values of n. Such tables are in a *cumulative* form. In other words, the probability value read from Table A-1 in the appendix will be the probability of getting r successes *or fewer*. A comparison of the cumulative probabilities versus the "point" or single probabilities is shown in Table 6-1 for $n = 5$ and $p = .60$ (the salesman's situation). We find the probabilities in the cumulative column merely by adding the probabilities from the point probability column.

The salesman in situation 1 may hand out special brochures to those new clients who have not already heard of his product. If he has two brochures with him, what are the chances that he will have enough for all five calls? This would be the same as asking: What is

TABLE 6-1.   Cumulative versus Point
Probabilities for the Binomial Distribution ($n = 5$, $p = .60$)

No. of Successes (r)	Point Probability (exactly r)	Cumulative Probability (r or fewer)
0	.0102	.0102
1	.0768	.0870
2	.2304	.3174
3	.3456	.6630
4	.2592	.9922
5	.0778	1.0000
	1.0000	

the probability that there will be two *or fewer* of the clients who have not heard of his product? We could look in Table A-1 and find that

$$\Pr(r \leq 2 | n = 5, q = .40) = .6826$$

Notice that we worked with $q = .40$, the probability of not having heard of the product, instead of $p = .60$, the probability of having heard of the product. An examination of Table A-1 in the appendix will show that the highest value of p tabulated is $p = .50$ because no higher value of p is needed. When p is higher than .50, we work with a corresponding number of failures because q would then have to be less than .50. The comparison of successes and failures is shown in Table 6-2.

TABLE 6-2

No. of Successes	---Is the Same Event as---	No. of Failures
0		5
1		4
2		3
3		2
4		1
5		0

In other words, the probability of zero successes is exactly the same as the probability of five failures. Therefore if the probability of a success is greater than .50, we merely work with the complementary number of failures where the probability of a failure is $1 - p$.

To find the probability of exactly r successes from the cumulative table, merely calculate

$$\Pr(\text{exactly } r \text{ successes} \,|\, n, p)$$

$$= \Pr(r \text{ or fewer successes} \,|\, n, p)$$
$$- \Pr(r - 1 \text{ or fewer successes} \,|\, n, p)$$

At this point you should use the tables to verify the probabilities calculated by the formula for situations 1 and 2.

To illustrate some tricks of working with a binomial table, suppose the salesman wanted to know the probability that two or fewer clients *have* heard of his product. We cannot work with the "have-heards" because $p = .60$. Therefore, by referring to Table 6-2, we see that two or fewer "have-heards" is exactly the same event as three or more "have-not-heards" with $q = .40$. When we go to look up

$$\Pr(r \geqslant 3 \,|\, n = 5, p = .40)$$

we realize that these tables give us the probability of r or less — not r or more. We must therefore make one final conversion. Notice that the events five "have-not-heards," four "have-not-heards," three "have-not-heards," two "have-not-heards," one "have-not-heard," and zero "have-not-heards" are mutually exclusive and collectively exhaustive. Therefore the total of the probabilities of the events equals 1. In other words, one of them and only one must happen. To find $\Pr(r \geqslant 3 \,|\, n = 5, p = .40)$, we can find $\Pr(r \leqslant 2 \,|\, n = 5, p = .40)$ and subtract it from the total, 1. In summary:

$$\Pr(r \leqslant 2 \,|\, n = 5, p = .60) = \Pr(r \geqslant 3 \,|\, n = 5, q = .40)$$

$$= 1 - \Pr(r \leqslant 2 \,|\, n = 5, p = .40)$$

$$= 1 - .6826$$

$$= .3174$$

Don't worry—a little practice will do wonders for you. Always draw a chart like Table 6-2 — it helps.

The binomial distribution can typically be applied when we are

sampling with replacement or sampling from a relatively large population so that the probability changes by a trivial amount each trial.
Sampling with replacement means that every time we examine an
object from the population we replace it (put it back) in the population so that it has the same chance of being chosen again on subsequent trials. When we are sampling from a small population without
replacement, we can no longer use the binomial distribution. We
can, however, often use the hypergeometric distribution.

THE HYPERGEOMETRIC DISTRIBUTION

The hypergeometric distribution commonly occurs in business
situations. Suppose that in situation 1 involving the salesman and
five new accounts, the salesman sent brochures to three of the five
accounts but lost his file and cannot remember which of the three
accounts received them. What are the chances that in the first four
calls he finds exactly two accounts that have already received the
brochures? We can solve this problem using a hypergeometric formula because our population is small (only these five accounts make
up the population) and because we are sampling without replacement
(once he has called an account he knows whether or not the account
received a brochure). The hypergeometric fomula is

SO&B 6-2: $$\Pr(r \text{ successes}) = \frac{{}_xC_r \; {}_{N-x}C_{n-r}}{{}_NC_n}$$

where N = number of items in the population
$\quad\quad x$ = number of items having the desired characteristic
$\quad\quad\quad$ (success)
$\quad\quad n$ = total number of items to be sampled

(Refer to Chapter 5 for a review of ${}_nC_r$ notation.) In our example
above,

$\quad\quad\quad\quad N = 5$ $\quad$ (total number of new accounts)

$\quad\quad\quad\quad x = 3$ $\quad$ (number of accounts that have received a
$\quad\quad\quad\quad\quad\quad\quad\quad\quad$ brochure)

$\quad\quad\quad\quad n = 4$ $\quad$ (we are concerned about the first four
$\quad\quad\quad\quad\quad\quad\quad\quad\quad$ accounts)

Therefore, using SO&B 6-2,

$$Pr(2 \text{ received brochure}) = \frac{{}_3C_2 \, {}_2C_2}{{}_5C_4}$$

$$= \frac{3!/2!1! \cdot 2!/2!0!}{5!/4!1!}$$

$$= \frac{3}{5}$$

Suppose a company has a typing pool of ten secretaries to handle overloads in various departments. Six of the secretaries can take shorthand and four cannot. A department manager puts in an order for three girls from the typing pool. However, he forgets to stipulate that one girl must take shorthand and the other two need not. What are the chances that he will get what he wants by random choice? Again this is a small population and the selection is done without replacement. Using the hypergeometric formula (SO&B 6-2), we find that

$$Pr(1 \text{ shorthand, 2 no shorthand}) = \frac{{}_6C_1 \, {}_4C_2}{{}_{10}C_3}$$

$$= \frac{6!/5!1! \cdot 4!/2!2!}{10!/7!3!}$$

$$= \frac{6 \cdot 6}{120}$$

$$= .30$$

THE POISSON DISTRIBUTION

A third distribution that commonly appears in business and industry is the *Poisson distribution*. The Poisson distribution has a number of uses, but the one we will talk about is one in which the number of possible successes or failures is very large or uncountable. For instance, the paint on a refrigerator door has an uncountable number of possibilities for blemishes. But if certain characteristics can be determined and we know the average number of defects per door, we can find the probabilities of various numbers of blemishes by applying the Poisson distribution. The Poisson can often be used when we are working with intervals of time (such as the number of phone calls on a switchboard in a given time interval), a surface area (such as the paint blemishes on a finished automobile fender), and intervals of length (such as the number of defects in a length of extruded plastic hose). Now let us formalize these concepts. There are three assumptions behind the Poisson distribution:

1. The number of successes in any interval or space is independent of the number of successes in other intervals or spaces.

2. The probability that a success will occur in any small interval or space is proportional to the size of that interval or space.

3. In an extremely small interval or space the probability of two successes occurring is negligible.

If these three characteristics apply to a situation, then we can determine the probability of the number of successes by applying the Poisson distribution. If the average number of successes in a given interval (space) is μ and the three preceding characteristics hold, then the probability of r successes occurring in that interval is given by

$$\Pr(x;\mu) = \frac{e^{-\mu}\mu^x}{x!}$$

Fortunately, the values in the Poisson distribution for various μ's and x's have been tabulated. Table A-2 in the appendix presents the *cumulative* values of x for a limited number of values of μ in a Poisson distribution.

Suppose that a manufacturer of 12×12 inch tiles is willing to accept as true the three characteristics required for the Poisson distribution. He knows that there is an average of one blemish per tile. Tiles that are blemish-free can be sold at a premium price. What percentage of his production should he expect to be able to sell at a premium? Since

$$\Pr(x = 0 | \mu = 1) = .36788$$

he would expect approximately 36.8 percent of his production to bring a premium price.

Now suppose that the manufacturer also produces in the same process a 12×24 tile. This larger tile has exactly twice the area of the 12×12 tile. By the second characteristic required, we know that the probabilities associated with a Poisson distribution are proportional to the interval in space. We find the probabilities associated with the tile that is doubled in size by doubling the value of μ: μ (for 12×12) = 1, which doubled gives μ (for 12×24) = $1 \times 2 = 2$. Then the proportion of the large tiles expected to be free of defects would be found by

$$\Pr(x = 0 \,|\, \mu = 2) = .13534$$

Now take another example. There is an average of two people arriving at a supermarket checkout stand each five minutes. The checker can handle three people per five-minute interval. If more than three people arrive, a queue or line begins to develop. In a five-minute interval, what is the probability that a queue will begin to develop? We calculate

$$\Pr(x \geq 4 \,|\, \mu = 2) = 1 - \Pr(r \leq 3 \,|\, \mu = 2)$$
$$= 1 - .85712$$
$$= .14288$$

Depending on the type of store, this may be felt by management to have a rather high chance of a line or queue developing and customer dissatisfaction resulting. The manager may want to speed up the checkers or install another checkout stand.

The Poisson distribution is used quite commonly in *queuing problems* involving arrival of customers for service. Queuing theory itself is an entirely different and complex subject.

DISTRIBUTIONS WITH CONTINUOUS VARIABLES

Thus far we have concerned ourselves with discrete variables that can take on only a countable number of values. We have been counting such items as number of defects, number of people, and so on, which cannot be fractions. But other variables can take on *any* value — the speed of a car, the amount of electricity consumed, the amount of time taken to complete a task, the weight of a machine-filled container. The latter variables, those that can take on any value, are *continuous* variables. The binomial, hypergeometric, and Poisson distributions handle discrete variables. We must now examine one of the distributions for continuous variables.

THE NORMAL DISTRIBUTION

The normal distribution is probably the most important distribution in the study and use of statistics. It has the characteristics of being symmetric about its mean μ and of having almost all (99 percent) of its values within ± 3 standard deviations of the mean. When graphed, the normal distribution is the familiar bell-shaped curve. The equation for the normal distribution is

$$f(x) = \frac{1}{\sqrt{2\pi}\sigma} e^{-\frac{1}{2}\left(\frac{x-\mu}{\sigma}\right)^2}$$

where x is the variable and μ and σ are the two parameters. There is a whole family of normal distributions, one for each different μ and each different σ. Before you get hypertense over the complexity of this formula and this discussion, though, rest assured that statisticians, being as basically lazy as any normal person, have a way of simplifying the whole procedure.

If a variable x is normally distributed, then the *standard normal deviate*,

SO&B 6-3: $$z = \frac{x - \mu}{\sigma}$$

is also normally distributed. The standard normal deviate, in fact, is normally distributed with a mean μ equal to zero and a standard deviation σ equal to 1. This means that any normal distribution can be reduced to one in which $\mu = 0$ and $\sigma = 1$. We can then make up a table of values for this one special case and bring the whole ball game under control. Such a table appears in the appendix as Table A-3.

In Table A-3 the values are the proportion of a normal distribution that lies between the mean and a certain number of standard deviations from the mean. Looking back at SO&B 6-3,

$$z = \frac{x - \mu}{\sigma}$$

you can see that z is really the number of standard deviations that the observation x is from the mean μ. In Table A-3 the left-hand column is the value of z to the nearest tenth and the column headings are an extension of this z value to the nearest one-hundredth. If we wanted to know the proportion of the curve that is between the mean and the mean plus *one* standard deviation, we would look up in Table A-3 the value associated with $z = 1$, which is .3413. Figure 6-1 is a schematic of this procedure.

FIGURE 6-1. The Normal Curve

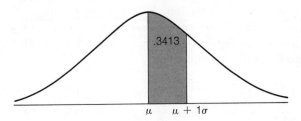

If we wanted to know the proportion of the curve that is between the mean and the mean plus 1.05 standard deviations, we would look up in Table A-3 the value associated with $z = 1.05$. We find this by coming down the left column to 1.0, then going across that row to the column headed ".05." The appropriate value is .3531.

Now let's see if we can get this all together by going through an example. Our company is making building-bricks from the solid waste in an urban dump. The weights of these bricks are approximately normally distributed with an average of 5 pounds and a standard deviation of .20 pounds. A customer desires bricks that are between 5 and 5.2 pounds. What proportion of our stock would meet his needs? Figure 6-2 shows the problem. We know that $\mu = 5$, so we are looking for the area between the mean (5) and $x = 5.2$. We calculate the standard normal deviate for $x = 5.2$ as

$$z = \frac{5.2 - 5}{.20}$$

$$= 1$$

FIGURE 6-2. Normal Curve ($\mu = 5$, $\sigma = .20$)

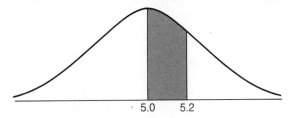

This tells us that the value 5.2 is one standard deviation from the mean. We find the proportion associated with $z = 1$ in Table A-3 to be .3413 (the same numbers we worked with in the explanation). Hence we would expect 34.13 percent of our stock to meet the customer's needs.

If the customer wants bricks that weigh between 5 and 5.21 pounds, we approach the problem in the same manner. We know that the mean μ equals 5, so we are looking for the area or proportion of the normal curve between the mean (5) and $x = 5.21$. Calculating the z value for 5.21, we find that

$$z = \frac{5.21 - 5}{.20}$$

$$= 1.05$$

This tells us that the value 5.21 is 1.05 standard deviations from the mean. Looking in the proper table we find that .3531 of our inventory would be expected to fit this customer's needs.

FIGURE 6-3. Normal Curve ($\mu = 5$, $\sigma = .20$)

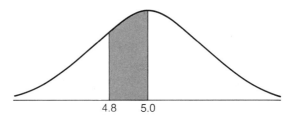

4.8   5.0

How would we handle the problem if we were looking for the proportion of bricks that weighed between 4.80 and 5 pounds? We would be looking for the area indicated in Figure 6-3. The only difference between this figure and Figure 6-2 (where we wanted bricks between 5 and 5.20 pounds) is that here the area lies to the left of the mean instead of the right. When we calculate the z value,

$$z = \frac{x - \mu}{\sigma}$$

$$= \frac{4.80 - 5.00}{.20}$$

$$= -1.00$$

we find that we have −1 instead of +1. Remember, though, that we said the normal curve is symmetric. Therefore there is the same area between the mean and −1σ as there is between the mean and +1σ. The proportion of bricks that weigh between 4.80 and 5 pounds is .3413, the same as the proportion of bricks that weigh between 5 and 5.20 pounds. Similarly, the proportion of bricks that weigh between 5 and 4.79 pounds is .3531, where the z value is

$$z = \frac{4.79 - 5}{.20}$$

$$= -1.05$$

Carrying this a step further, let's look at the proportion of bricks that weigh between 4.80 and 5.20 pounds. We are looking for the area indicated in Figure 6-4. Whenever we are working with values

FIGURE 6-4. Normal Curve ($\mu = 5$, $\sigma = .20$)

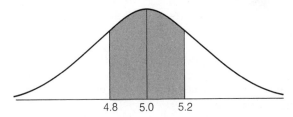

in a normal curve we must always work with their distance from the mean when we are using tables in the form of Table A-3. That means we must find the proportion of bricks between 5.20 and 5 (.3413) and the proportion of bricks between 4.80 and 5 (.3413). Adding these two proportions together, we find that

$$.3413 + .3413 = .6826$$

Thus 68.26 percent of our bricks will weigh between 4.80 and 5.20 pounds.

Suppose that Mabel Metcher comes into our place of business and wants a brick that weighs between 5.5 and 5.6 pounds. (She wants to keep it in her purse to discourage her more ardent customers.) What is the probability that if she randomly chooses a brick, it will be the proper weight? In this problem we are looking for the area shown in

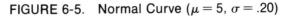

FIGURE 6-5.   Normal Curve ($\mu = 5$, $\sigma = .20$)

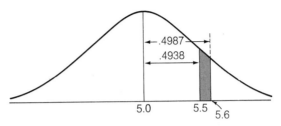

Figure 6-5. Again, with the tables we are using we must put the values relative to the mean. We will find the proportion or area between the mean and 5.6; then we will subtract from that the proportion or area between the mean and 5.5. This will leave us the area between 5.5 and 5.6:

$$z = \frac{5.6 - 5}{.20} = 3.00 \qquad \text{area for } 3.00 = .4987$$

$$z = \frac{5.5 - 5}{.20} = 2.50 \qquad \text{area for } 2.50 = .4938$$

$$\Pr(5.60 \leqslant x \leqslant 5.50) = .4987 - .4938 = .0049$$

We can see that poor Mabel has a pretty slim chance of finding a brick of the proper weight.

One of our brick salesmen, also versed in statistics, recalculates the whole problem and reaffirms that Mabel has a slim chance of finding the brick she wants. The salesman has recently returned from a convention on selling bricks, so he immediately tries to switch her to a different purchase (he gets a commission for each brick he sells). He convinces her that carrying around a brick that heavy may be tiring and so she should be looking for a brick that is 5.6 pounds or less. The probability of selecting such a brick randomly is shown schematically by the shaded area in Figure 6-6.

From a previous calculation we know that the proportion of weights between 5.6 and 5 pounds is .4897. We also know that one-half the bricks or 50 percent weigh less than 5 pounds (are less than the mean). Therefore

$$\Pr(\text{brick weighs less than 5.6 lb}) = .4987 + .5000$$

$$= .9987$$

FIGURE 6-6.   Normal Curve ($\mu = 5$, $\sigma = .20$)

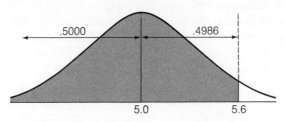

.5000   .4986

5.0   5.6

As a word of advice, I strongly suggest that when you are working with the normal distribution, you always draw a sketch — like the ones shown in Figures 6-1 through 6-6 — to help you keep track of the logic of the problem you're working on.

The normal curve is an extremely useful tool that gains its leverage and usefulness from three different areas:

1.   Many things seem to be approximately normally distributed by nature.

2.   The normal distribution can be used to approximate the binomial distribution when the sample size is large.

3.   The central limit theorem tells us that the distribution of means is approximately normally distributed.

We have just gone through an example of item 1. Item 3 is covered in a subsequent chapter (don't worry about it now). We will now look at item 2.

THE NORMAL APPROXIMATION TO THE BINOMIAL

The three assumptions or determining characteristics of the binomial distribution were listed as:

1.   There are only two possible outcomes of each experiment or trial.

2.   The probability of a success remains constant from trial to trial.

3.   The trials are independent of each other.

If these characteristics are satisfied, the number of successes r out of n trials follows a binomial distribution.

We often encounter the situation in which the binomial distribution can be assumed, but n, the number of trials, is so large that the calculations become tedious and a complete table of binomial values is not available for that size n. In these situations we can (within reasonable ranges) use the normal distribution to find a close approximation to the true binomial distribution. The qualifying phrase "within reasonable ranges" refers to the size of n and the size of p. Some statisticians recommend that $n \cdot p$ and $n \cdot q$ should *both* be equal to 5 or greater before the normal approximation to the binomial can be applied. We will follow that rough rule of thumb.

To use the normal approximation technique, we can follow either of two approaches. Both give the same result. Approach 1—using frequency—comprises three steps:

1. Let $\mu = nP$ (where P is the probability of success on any trial) and let $\sigma = \sqrt{nPQ}$.

2. Calculate the z value:

SO&B 6-4:
$$z = \frac{x - \mu \pm \frac{1}{2}}{\sqrt{nPQ}}$$

where x is the frequency or number of occurrences with which we are concerned. We will see, in the examples that follow, when to add or subtract the continuity correction factor of $\frac{1}{2}$.

3. Find the associated probability in the table of normal values (Table A-3).

Approach 2—using proportions—also comprises three steps:

1. Let $\mu = P$ (where P is the probability of success on any trial) and let $\sigma = \sqrt{PQ/n}$.

2. Calculate the z value:

SO&B 6-5:
$$z = \frac{p - \mu \pm 1/(2n)}{\sqrt{PQ/n}}$$

where p is the proportion with which we are concerned. We will see, in the examples that follow, when to add or subtract the continuity correction factor of $1/2n$.

3. Find the associated probability in the table of normal values (Table A-3).

Example 1: The police chief in a large city is cracking down on illegal gambling. He is relying on informants for addresses, but the information is inaccurate and only 60 percent of his raids are at the right place. Yesterday, for instance, he and his vice squad, axes in hand, hacked their way into three illegal dice games, one bingo game for senior citizens, one monthly meeting of the Rose and Garden Club, and one game of Monopoly being played by four children.

" WHAT'S THE PROBABILITY OF DOING THIS AGAIN, CHIEF ? "

Besides all the bad publicity, the chief has to repair the chopped-up doors at the places he raided in error, and this is hurting his annual budget. (In addition he had to give each of the four children at the Monopoly game a permanent Get Out of Jail Free card.) His experience tells him that the probability of breaking into the right address is .60. He also believes that each raid is independent of all others. It is a large city with a high number of illegal gambling games.

Next week the chief plans to make 50 raids. His quota of successful raids is 28. He might also say that his quota for the 50 raids is 56 percent ($\frac{28}{50} \cdot 100$) successes. Both statements say the same thing. If he is truly making 60 percent successful raids, what are the chances that he will meet his quota? This is a binomial problem and we will apply the normal approximation. First, using approach 1 with frequency, we would calculate

$$\mu = nP$$
$$= 50 \cdot .60 \quad \text{(sample size} = 50;$$
$$\qquad\qquad\qquad\text{probability of success} = .60)$$
$$= 30$$

$$\sigma = \sqrt{nPQ}$$
$$= \sqrt{50(.60)(.40)}$$
$$= \sqrt{12}$$
$$= 3.46$$

We are looking for $\Pr(r \geq 28 \,|\, n = 50, P = .60)$. Figure 6-7 illustrates that although the binomial distribution is in terms of whole numbers (27 successes, 28 successes, 29 successes, and so forth), we are trying to approximate it with the continuous normal curve. Therefore, instead of the observation "28 successes" being just the point 28, it is the rectangular area from 27.5 to 28.5. Similarly, 29 is the area between 28.5 and 29.5, and so on for the whole-number observations.

We are looking for the probability that there will be 28 or more successes. From Figure 6-7 we see that since 28 actually extends to 27.5 in the normal approximation, we

FIGURE 6-7. Normal Approximation
to Binomial ($\mu = 30$, $\sigma = 3.46$)

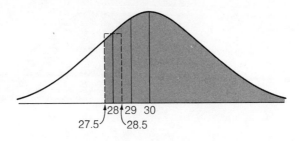

would *subtract* $\frac{1}{2}$ for the continuity correction factor in calculating the z value:

$$z = \frac{27.5 - 30}{3.46}$$

$$= \frac{-2.5}{3.46}$$

$$= -.72$$

From Table A-3 we find that .2642 proportion of a normal distribution is between the mean and the mean minus $.72\sigma$. There is also .5000 proportion of the distribution to the right of the mean 50:

$$\Pr(r \geqslant 28 \,|\, n = 50,\ P = .60) = .2642 + .5000$$

$$= .7642$$

For the same problem with approach 2 (using proportions), we would calculate

$$\mu = P$$

$$= .60$$

$$\sigma = \sqrt{\frac{PQ}{n}}$$

$$= \sqrt{\frac{(.60)(.40)}{50}}$$

$$= .069$$

FIGURE 6-8.   Normal Curve ($\mu = .60$,

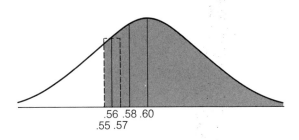

.56 .58 .60
.55 .57

We are looking for $\Pr(p \geq .56 \,|\, n = 50, P = .60)$. Figure 6-8 illustrates the approximation again of using the continuous normal curve to approximate the discrete binomial. Notice that in our sample of 50 we can have .60 proportion of successes ($\frac{30}{50}$), we can have .58 proportion of successes ($\frac{29}{50}$), or we can have .56 proportion of successes ($\frac{28}{50}$). But there's no way of having a proportion of successes of .59, .57, and so on. Thus, as in approach 1, each possible proportion of successes is a rectangular area rather than just a point or a line. The proportion .56, for instance, is an area from .55 to .57. In general, each proportion p is an area of $p \pm 1/2n$.

We are looking for the probability that the proportion of successes will be .56 or greater. From Figure 6-8 we see that since .56 actually extends to .55 in the normal approximation, we would *subtract* $1/2n$ or $1/100$ for the continuity correction factor in calculating the z value:

$$z = \frac{.55 - .60}{.069}$$

$$= -\frac{.05}{.069}$$

$$= -.72$$

This is exactly the same result we got using approach 1 (as it must be) and

$$\Pr(p \geq .56 \,|\, n = 50, P = .60) = .2642 + .5000$$

$$= .7642$$

Example 2: A company has been guaranteed that the television show they are sponsoring will be watched by 30 percent of the people. The marketing department has been assigned the task of making a telephone survey: they will randomly select 100 names, call those names during the television show, and ascertain whether the people are watching it or not. The executive of the sponsoring company will be extremely happy if 40 percent or more are found to be watching his program. If there are actually 30 percent watching the program, what is the probability that the sample will show 40 percent or more? Let's begin with approach 1 (frequencies):

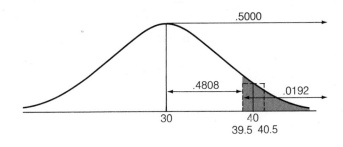

$$z = \frac{39.5 - 30}{\sqrt{100(.30)(.70)}}$$

$$= \frac{9.5}{\sqrt{21}}$$

$$= 2.07$$

$$\Pr(30 \leqslant x < 40) = .4808$$

$$\Pr(x \geqslant 40) = .5000 - \Pr(30 \leqslant x < 40)$$

$$= .5000 - .4808$$

$$= .0192$$

There is approximately a 2 percent chance of getting a sample value of 40 or greater.

Now let's use approach 2 (proportions):

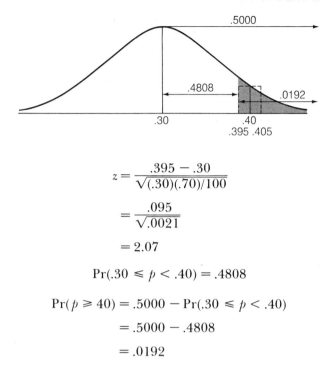

$$z = \frac{.395 - .30}{\sqrt{(.30)(.70)/100}}$$

$$= \frac{.095}{\sqrt{.0021}}$$

$$= 2.07$$

$$\mathrm{Pr}(.30 \leqslant p < .40) = .4808$$

$$\mathrm{Pr}(p \geqslant 40) = .5000 - \mathrm{Pr}(.30 \leqslant p < .40)$$

$$= .5000 - .4808$$

$$= .0192$$

There is approximately a 2 percent chance of getting a sample value of .40 or greater.

There are two plausible translations of this result:

1.   If 30 percent of the people were watching this show, the producers had better not count on making the sponsor happy.

2.   If the sample result did come back at 40 percent or more, there is a good chance that more than 30 percent of the people were watching this show.

AUTOPSY

We have looked at a few of the most practical and useful probability distributions. There are many probability distributions available to the analyst. Learning how to find probabilities from a probability distribution is a necessary step, but you should remember that the most important part is recognizing that a particular situation may fit a particular probability distribution. Some of these situations can be recognized by their characteristics; others may, after careful examination, be assumed to follow a particular distribution approximately.

SUMMARY

Binomial Distribution

SO&B 6-1: $\Pr(r \text{ successes} | n \text{ trials}, p) = {}_nC_r p^r q^{n-r}$

(values are tabled in appendix)

Required characteristics:

1. Only two possible outcomes.

2. The probability of a success (p) remains constant from trial to trial.

3. The n trials are independent of each other.

Hypergeometric Distribution (used when sampling from a small group without replacement):

SO&B 6-2: $\Pr(r \text{ successes}) = \dfrac{{}_xC_r \; {}_{N-x}C_{n-r}}{{}_NC_n}$

Poisson Distribution:

$$\Pr(x \text{ successes} | \mu) = \frac{e^{-\mu}\mu^x}{x!}$$ (values are tabled in appendix)

Required characteristics:

1. Each interval is independent of all other intervals.

2. The probability that a success will occur in any interval is proportional to the size of that interval.

3. The probability of two successes occurring in an extremely small interval is negligible.

Normal Distribution

Characteristics:

      1. Bell-shaped

      2. Symmetric about the mean

      3. Continuous

      4. Most important distribution in statistical analysis. Values are tabled for standard normal deviate:

SO&B 6-3:
$$z = \frac{x - \mu}{\sigma}$$

      5. Can be used to approximate the binomial distribution by using frequencies:

SO&B 6-4:
$$z = \frac{x - \mu \pm \frac{1}{2}}{\sqrt{nPQ}}$$

or by using proportions:

SO&B 6-5:
$$z = \frac{p - \mu \pm 1/(2n)}{\sqrt{PQ/n}}$$

PROBLEMS

1. Describe a situation that would require using the binomial distribution.

2. Describe a situation that would require using the hypergeometric distribution.

3. Describe a situation that would require using the Poisson distribution.

4. Describe a situation that would require using the normal distribution.

5. Evaluate the following using the binomial tables:

      a. $\Pr(r = 0 \mid n = 5,\ P = .20)$

      b. $\Pr(r \leqslant 1 \mid n = 5,\ P = .20)$

      c. $\Pr(r \leqslant 4 \mid n = 7,\ P = .40)$

$\quad\quad$ d. $\;$ $\Pr(r = 7 | n = 7, P = .40)$

$\quad\quad$ e. $\;$ $\Pr(r \leqslant 6 | n = 10, P = .50)$

$\quad\quad$ f. $\;$ $\Pr(r \geqslant 2 | n = 10, P = .60)$

$\quad\quad$ g. $\;$ $\Pr(r \geqslant 3 | n = 5, P = .70)$

$\quad\quad$ h. $\;$ $\Pr(r \leqslant 1 | n = 5, P = .05)$

$\quad\quad$ i. $\;$ $\Pr(r = 2 | n = 4, P = .10)$

$\quad\quad$ j. $\;$ $\Pr(1 \leqslant r \leqslant 5 | n = 7, P = .40)$

Answers: $\;$ (a) .3277 $\quad\quad$ (b) .7373 $\quad\quad$ (c) .9037 $\quad\quad$ (d) .0016 $\quad\quad$ (e) .8281
(f) .9983 $\quad\quad$ (g) .8369 $\quad\quad$ (h) .9774 $\quad\quad$ (i) .0486 $\quad\quad$ (j) .8757

6. $\;$ Evaluate the following using the Poisson tables:

$\quad\quad$ a. $\;$ $\Pr(r = 0 | \mu = 1)$

$\quad\quad$ b. $\;$ $\Pr(r = 1 | \mu = 1)$

$\quad\quad$ c. $\;$ $\Pr(r \leqslant 2 | \mu = .4)$

$\quad\quad$ d. $\;$ $\Pr(r \leqslant 4 | \mu = 4)$

$\quad\quad$ e. $\;$ $\Pr(r \geqslant 3 | \mu = 4)$

$\quad\quad$ f. $\;$ $\Pr(r \geqslant 2 | \mu = .5)$

$\quad\quad$ g. $\;$ $\Pr(r \leqslant 7 | \mu = 6)$

$\quad\quad$ h. $\;$ $\Pr(2 \leqslant r \leqslant 7 | \mu = 6)$

Answers: (a) .3679 $\quad\quad$ (b) .3679 $\quad\quad$ (c) .9921 $\quad\quad$ (d) .6288 $\quad\quad$ (e) .7619
(f) .0902 $\quad\quad$ (g) .7440 $\quad\quad$ (h) .7267

7. $\;$ Verify the following using the normal tables (draw a sketch of each).

If $\mu = 0$, $\sigma = 1$ find:

$\quad\quad$ a. $\;$ $\Pr(x \geqslant 0) = .5000$

$\quad\quad$ b. $\;$ $\Pr(0 \leqslant x \leqslant 1.5) = .4332$

$\quad\quad$ c. $\;$ $\Pr(x \leqslant 1.5) = .9332$

$\quad\quad$ d. $\;$ $\Pr(x \leqslant -1.5) = .0668$

If $\mu = 10$, $\sigma = 2$ find:

$\quad\quad$ e. $\;$ $\Pr(x \geqslant 10) = .5000$

$\quad\quad$ f. $\;$ $\Pr(10 \leqslant x \leqslant 13) = .4332$

$\quad\quad$ g. $\;$ $\Pr(x \leqslant 13) = .9332$

$\quad\quad$ h. $\;$ $\Pr(x \leqslant 7) = .0668$

8. A salesman is trying to get a service station operator to buy an inventory of first aid kits. He claims that 25 percent of the station's customers will buy one. The service station operator states that he will buy an inventory of the kits if at least one of the next four customers buys a kit.

> *a.* If the salesman's claim is true, what is the probability that he will sell an inventory to the service station operator?
>
> *b.* What is the probability that the service station operator will buy the inventory if, in fact, the salesman has exaggerated and only 10 percent of the customers will buy a kit? (*Hint:* Compare this situation to the required characteristics of the binomial distribution.)

9. The sales manager of a large automobile dealership estimates that 90 percent of the new cars delivered have no defect that will require the car to be brought back for repair. He has just sold an important commercial customer a fleet of 15 cars.

> *a.* What is the probability that none of the cars will need repair?
>
> *b.* What is the probability that five or more cars will need repair?

10. A union organizer claims that 60 percent of the employees of a large company want to join the union. In a random sample of 80 workers, 45 were in favor of the union. Would you believe the organizer's claim? Calculate the probability of getting 45 or fewer in favor of the union by utilizing the normal approximation to the binomial.

11. The number of typographical errors in a certain newspaper is known to follow a Poisson distribution with an average of one error per page.

> *a.* On a randomly drawn page, what is the probability of having more than two errors?
>
> *b.* What proportion of the pages would you expect to be error-free?
>
> *c.* What is the probability that two randomly drawn pages will be error-free?

12. A manufacturer of copper tubing knows that one of his processes (one-inch tubing) follows a Poisson distribution with an average

of one defect per 60 feet of tubing. A standard length of one-inch tubing is six feet. The company marks as reject and reprocesses any six-foot length that has more than two defects. What proportion of their six-foot-length production is rejected?

13.   A magazine claims that it is read by 80 percent of the business executives in the country. Suppose this claim is true.

> *a.*   What is the probability of finding fewer than 12 readers in a sample of 20?
>
> *b.*   What is the probability of finding fewer than 90 readers in a sample of 120?

14.   A highway construction firm knows that on a certain project, a dump truck will have a tire failure once every 20 working days. They have ascertained that tire failures follow a Poisson distribution. If they are operating a fleet of ten trucks, how many tires should the repair crew stock each day so that they will have less than one chance in a hundred of not having enough tires to make the necessary repairs?

15.   The bursting strength of a certain hydraulic hose is known to average 500 pounds with a standard deviation of 5 pounds. It is also approximately normally distributed.

> *a.*   What proportion of these hoses will burst only at more than 512 pounds of pressure?
>
> *b.*   What proportion of these hoses will burst at less than 490 pounds?

16.   An auditor has pulled the ten accounts of clients that are listed as bad debts. But unbeknownst to the auditor, two of the bad accounts are actually phony names the office manager is using to embezzle funds.

> *a.*   If the auditor randomly selects three of the ten accounts, what is the probability that he will not get a phony?
>
> *b.*   If the auditor randomly selects three accounts, what is the probability that he will get at least one phony?
>
> *c.*   If the auditor samples five instead of three accounts, what is the probability that he will not get a phony account? (*Hint:* The auditor is sampling without replacement for a small population.)

17.   Weird Willy, the statistician, buys three cans of mustard-soaked sunflower seeds and two cans of pickled squash. He intends to open one can for each dinner, but his two-year-old child rips the labels off all five cans. For the next three dinners, what is the probability that Willy will eat exactly one can of mustard-soaked sunflower seeds?

18.   A nurse rushing back to her station drops the six hypodermic needles she is carrying. Two of them have already been used and are therefore not sterile. Suppose she selects needles randomly for the next three patients.

> *a.*   What is the probability that all the patients will get an unused needle?
>
> *b.*   What is the probability that exactly one patient will get a used needle?

19.   A machinist making precision parts turns out 20 percent defective, or reject, parts. Each part is independent of the others he makes. Suddenly a rush order for 15 parts comes to him. He has time to make 20 parts. What is the probability that he can fill the order?

20.   The heat in an office is known to follow a normal distribution with a mean of 72° and a standard deviation of 1.5°.

> *a.*   What percentage of the time will the office be hotter than 74°?
>
> *b.*   What percentage of the time will the office be between 69° and 73°?

21.   Suppose that 20 percent of a company's inventoried products are obsolete and that the company has a large number of products. An accountant randomly selects a sample of ten items. What is the probability that he will find one or less obsolete products? The accountant increases his sample size to 50. What is the probability that he will find ten percent or fewer obsolete products in his sample?

22.   The scores on a certain aptitude test are known to be normally distributed with a mean of 76.4 and a standard deviation of 8.2 points. Find the percentage of test takers you would expect to find having the following scores:

> *a.*   Greater than 84.7.
>
> *b.*   Less than 85.0.
>
> *c.*   Between 70 and 80.

SELECTED REFERENCES

Bryant, Edward C. *Statistical Analysis.* Revised edition. New York: McGraw-Hill, 1966.

Dixon, W. J. and F. J. Massey. *Introduction to Statistical Analysis.* Third Edition. New York: McGraw-Hill, 1969.

Guenther, William C. *Concepts of Statistical Inference.* Second edition. New York: McGraw-Hill, 1973. Especially Chapter 2.

Miller, Irwin and John E. Freund. *Probability and Statistics for Engineers.* Englewood Cliffs, N.J.: Prentice-Hall, 1965. Especially Chapters 3, 4.

Spurr, William A. and Charles P. Bonini. *Statistical Analysis for Business Decisions.* Revised edition. Homewood, Ill.: Irwin, 1973. Especially Chapter 6.

Tanur, Judith M., Frederick Mosteller, William H. Kruskal, Richard F. Link, Richard S. Pieters, and Gerald R. Rising. *Statistics: A Guide to the Unknown.* San Francisco: Holden-Day, 1972.

Yamane, Taro. *Statistics: An Introductory Analysis.* Third edition. New York: Harper & Row, 1973. Especially Chapter 6.

CHAPTER 7
EXPECTED VALUES AND PAYOFF TABLES

OBJECTIVES: After studying this chapter and working the problems, you should be able to:

Define the term expected value of a variable.

Explain how to use subjective probabilities in a payoff table.

Define the terms action *and* event.

Construct a payoff table.

Make a decision based on the payoff table.

Demonstrate the use of EVPI.

THUS FAR WE have dealt with expected values on an intuitive level by referring to such things as the proportion of defects we would "expect" to occur. The *expected value* is the average of the values we would get in conducting an experiment or trial exactly the same way many times. The expected value is the mean μ. The expected value of a variable x is denoted by $E(x)$. The formula for $E(x)$ in a discrete situation* is

SO&B 7-1: $$E(x) = \Sigma[x \cdot \mathrm{Pr}(x)]$$

* For continuous situations, calculus is required:

$$E(x) = \int_a^b x \cdot f(x)\, dx \qquad a \leqslant x \leqslant b$$

If we were flipping a coin and every time a head came up you lost $1 as opposed to winning $1 every time a tail came up, you would expect in the long run to break even. If we apply the concept of expected values and SO&B 7-1, this is exactly what we find. Let x be the gain or loss of $1 on each flip of the coin. Then if the coin comes up heads, $x = -\$1$; and if the coin comes up tails, $x = +\$1$. Our expected profit from the transaction is thus

$$E(x) = \Sigma[x \cdot Pr(x)]$$
$$= (-1 \cdot \tfrac{1}{2}) + (+1 \cdot \tfrac{1}{2})$$
$$= -\tfrac{1}{2} + \tfrac{1}{2}$$
$$= 0$$

So we would break even in the long run.

Notice that on any toss of the coin it is impossible to come up with the expected value of $0; the outcome is either $-$1 or $+$1. The expected value tells you how you would "expect" to average out in the long run. But the expected value is useful to us even when we are going to conduct only one trial. This can be seen intuitively by changing the preceding example slightly. Suppose that if a tail comes up you still receive $1, but if a head comes up you lose $10. You still have a 50 percent chance of making a profit of $1 on one toss of the coin, but you would not agree to these terms because "the odds are against you." Formally,

$$E(x) = (-10 \cdot \tfrac{1}{2}) + (1 \cdot \tfrac{1}{2})$$
$$= -5 + 1$$
$$= -\$4$$

In Problem 3 of Chapter 5 an auditor was looking at 1000 accounts receivable, some of which were past due. The days past due were as follows:

No. of Days Past Due (x)	No. of Accounts	Pr(x)
0	600	.60
30	250	.25
60	100	.10
90	50	.05
	1000	1.00

The expected value of the number of days past due is

$$E(x) = (0 \cdot .60) + (30 \cdot .25) + (60 \cdot .10) + (90 \cdot .05)$$
$$= 0 + 7.50 + 6.00 + 4.50$$
$$= 18$$

The expected number of days past due on any randomly drawn account is thus 18 days.

When we are told that the probability of rain is only 10 percent, we do not *expect* it to rain. However, 10 times out of 100 under identical conditions it will rain. We have all been, at some time, drenched with two inches of 10 percent.

Expected values can be useful guides in decision making. Suppose a manager is contemplating building a new retail outlet. He estimates that if the economy goes up he will make $50,000 on the deal, if the economy remains stable he will make $10,000, and if the economy drops he will lose $50,000. He then estimates that the probability that the economy will go up is .20, the probability that the economy will remain stable is .30, and the probability that the economy will drop is .50. (These are *subjective* probabilities.) If he were to calculate the expected value of the venture by using SO&B 7-1, he would find:

Event	Profit (x)	Pr(x)
Economy up	$50,000	.20
Economy stable	10,000	.30
Economy down	−50,000	.50

$$E(x) = (50,000 \cdot .20) + (10,000 \cdot .30) + (-50,000 \cdot .50)$$
$$= 10,000 + 3,000 - 25,000$$
$$= -12,000$$

This should give the manager a strong impulse not to build the new retail outlet, all other things being equal. When alternative actions are available, we can often use the concept of expected values in making *payoff tables* as an aid to decision making. Payoff tables can be helpful to the decision maker in certain situations in which alternative actions are available. In fact, for a problem to exist at all, there must be more than one reasonable course of action. If only one course of action exists, there is no problem . . . you merely take that course of action. Many people waste their time dwelling on "problems" that don't exist because they really have no alternative.

There are four components of a payoff table: actions, events or states of nature, probabilities, and the net payoff of all combinations of events and actions. The *actions* are the alternatives or choices the decision maker can choose or control. Imagine that you are the manager of a modeling agency. Besides your regularly scheduled business you get calls for models on a last-minute basis. However, you must tell your models on the previous day whether or not you want them to come to work. If they just sit around the office with no calls, you must pay them their $50 daily fee anyway. If you get a call for a model and you have one available, you receive $75, which means a profit of $25 after the daily fee is paid. To simplify the problem, suppose that there is no penalty for not being able to fill a request. The actions available to you are these: tell no extra models to report for work tomorrow, tell one extra model to report for work tomorrow, tell two extra models to report for work tomorrow, and so forth. You control the choice of action.

The *events* are sometimes called states of nature—the occurrences over which you have little or no control. The events in our example are these: you get no calls tomorrow for an extra model, you get one call tomorrow for an extra model, you get two calls tomorrow for an extra model, and so forth. You have no control over how many calls you will receive. The events that can happen must be mutually exclusive and collectively exhaustive, as are the events in our example.

The *probabilities* are the chances of each event occurring. These probabilities can be from a priori, empirical, or subjective sources. Since the events are mutually exclusive and collectively exhaustive, their probabilities must add up to 1. With the modeling agency we may be able to go back to past records and find that on 40 percent of the days no extra models were needed, that on 50 percent of the days one extra model was needed, that on 10 percent of the days two extra models were needed, and that there has never been a need for more than two extra models. These empirical probabilities will serve as the probabilities for our events.

The *payoffs* are the direct profit or loss from the combination of each action with each event. If you told one model to report for work and no calls came for extra models that day, for instance, your profit would be −$50 (a loss). If you told one model to report for work and one call came for an extra model, your profit would be $25. If you told no one to report for work and received calls for two models, your profit would be $0. Remember: The payoff is always the *direct* profit or loss, and the fact that we could have made a profit is immaterial. We had no revenue and no cash outflow, so the payoff is $0.

We now put the four components into a payoff table. The payoff table for our example is shown in Table 7-1.

TABLE 7-1.   Payoff Table for Modeling Agency

Event	Pr(Event)	Action (No. of Models to Work) 0	1	2
No models needed	.40	$0	-$50	-$100
1 model needed	.50	0	25	-25
2 models needed	.10	0	25	50
	EMV	$0	-$5	-$47.50

The bottom line of a payoff table is the *expected monetary value* (EMV) of each action. The EMV of each action is found by the usual expected-value formula: the sum of each payoff multiplied by its probability of occurrence. The EMV for the action "one model told to report for work" is

$$EMV(1) = (-\$50 \cdot .40) + (\$25 \cdot .50) + (\$25 \cdot .10)$$
$$= -\$20 + \$12.50 + \$2.50$$
$$= -\$5$$

Assuming that money has a linear value ($2 is twice as good as $1), we choose the action that has the best EMV. In this case we would choose the action "no models told to report for work" because we break even on that action, whereas for the other two actions we would lose an average of $5 per day or $47.50 per day, respectively.

Let's take another example. A company is considering buying abandoned gold mining property (they are speculating on the future free-market price of gold). They have three prospective mines (A, B, C) from which to choose, but they will invest in only one. Their detailed engineering reports on the quality of each mine have given them the data to make the estimates for a payoff table, shown in Table 7-2.

TABLE 7-2. Payoff Table for Mining Venture

Event (Stabilized Price Per Oz. of Gold)	Pr(Event) (Subjective)	Do Nothing	A	Buy Mine B	C
				Action (in $10,000s)	
$120	.20	0	−10	0	5
$110	.50	0	0	20	10
$100	.30	0	50	30	20
EMV		0	13	19	12

The company's decision would be to buy mine B because it has the highest EMV (all other things being equal). Once again the EMV (for buy mine A, for instance) was found by

$$EMV = (-10 \cdot .20) + (0 \cdot .50) + (50 \cdot .30)$$
$$= -2 + 0 + 15.0$$
$$= 13.0$$

THE EXPECTED VALUE OF PERFECT INFORMATION

A natural extension of a payoff table is the *expected value of perfect information* (EVPI). The EVPI can be used as a rough analysis of whether or not it is worthwhile to spend money for further information. Looking back at our example with the modeling agency, suppose that a gypsy is willing to predict, with guaranteed accuracy, the number of models we will need for each following day. What is that information worth to us? First we will reconstruct Table 7-1.

TABLE 7-1 (again). Payoff Table for Modeling Agency

Events	Pr(Event)	0	1	2
		Action (No. of Models to Work)		
No models needed	.40	$0	−$50	−$100
1 model needed	.50	0	25	−25
2 models needed	.10	0	25	50
EMV		0	−$5	−$47.50

If the gypsy does predict accurately over a long period of time, our probabilities tell us she will say "no models needed" 40 percent of the time. Each time she tells us this we will assign no models to work and make a profit of $0, as indicated by the arrow. Over a long period, 50 percent of the time we will be told "one model is needed." Each of these times we will maximize our profit by assigning one model to report for work and make $25. Similarly, we will make a maximum profit of $50 (indicated by arrow) by assigning two models to report for work when she tells us "two models will be needed." We expect this to happen 10 percent of the time. With the gypsy giving us perfect predictions, our expected daily profit will be

$$E(\text{daily profit with certain information})$$
$$= (\$0 \cdot .40) + (\$25 \cdot .50) + (\$50 \cdot .10)$$
$$= 0 + \$12.50 + \$5.00$$
$$= \$17.50$$

The best action available to us without this perfect information was "no models report for work," which will give us an average daily profit of $0. Hence if we can average $17.50 a day with perfect information and only $0 a day without perfect information, the EVPI must have a value of

$$\$17.50 - \$0 = \$17.50 \text{ a day}$$

In other words, we should be willing to pay the gypsy up to $17.50 daily for her prediction. In general,

SO&B 7-2:   EVPI = expected value under certainty

                      − EMV of best action under uncertainty

Suppose in the example involving the mining venture that the company is considering hiring an international consulting firm to predict the future price of gold. The fee for the firm would be $75,-000. Should they hire the firm? We will use EVPI to answer this question. First let's reconstruct Table 7-2.

From Table 7-2 (again) we see that if the consultant firm were to predict a price of $120 per ounce, the company would buy mine C for a $50,000 profit; if a $110 price were predicted, the company would buy mine B for an estimated $200,000 profit; if the consultants

TABLE 7-2 (again). Payoff Table for Mining Venture

Event (Stabilized Price Per Oz. of Gold)	Pr(Event)	Action (in $10,000s)			
		Do Nothing	A	Buy Mine B	C
$120	.20	0	−10	0	5
110	.50	0	0	20	10
100	.30	0	50	30	20
	EMV	0	13	19	12

predicted $100 per ounce, the company would purchase mine A for a $500,000 profit. The company has estimated the probabilities for the price levels of $120, $110, and $100 to be .20, .50, and .30, respectively. The expected value under a perfect prediction would then be

$$E(\text{profit with perfect information})$$
$$= (\$50,000 \cdot .20) + (\$200,000 \cdot .50) + (\$500,000 \cdot .30)$$
$$= \$10,000 + \$100,000 + \$150,000$$
$$= \$260,000$$

If the expected value of the venture is $260,000 with perfect information and the expected value of the venture (EMV of mine B) without perfect information is $190,000, then the EVPI (using SO&B 7-2) is

$$\text{EVPI} = \$260,000 - 190,000$$
$$= \$70,000$$

The consultant firm's fee of $75,000 exceeds the EVPI. Even if they gave the speculators perfect information, we would not "on the average" expect the information to pay for itself. In fact, we know they can't deliver perfect information. Therefore the speculators should not hire the consultant firm.

AUTOPSY

The use of expected values and payoff tables is a valuable aid for the decision maker. But remember that they should generally be used as an aid and not as the final decision point. Other factors often have to be weighed.

Although this discussion was based strictly on monetary values, you should be aware that if some other subjective value must be attached to a payoff, *utility* values can be used. For instance, your decision on whether or not to carry a raincoat on a rainy day depends on the utility value you associate with getting wet. For each person this value would be different.

You should also be aware that money often is not linear. For instance, you may be willing to devote ten years of your life to a project that will make you $1 million. But you probably would not think it is twice as good to spend 20 years of your life on a project that will make you $2 million. In the first place, it might take you a lifetime to spend $1 million. And in the second place, once you had the first million, the second million might not have any appeal to you.

The material covered in this chapter is only an introduction to the use of payoff tables and the evaluation of additional information. More material on this subject can be found in the Selected References.

SUMMARY

SO&B 7-1: $E(x) = \Sigma[x \cdot \Pr(x)]$

Payoff tables—an aid to decision making—are made up of:

> Actions
>
> Events
>
> Probabilities
>
> Payoffs

SO&B 7-2:   EVPI = expected value under certainty

                   − EMV of best action under uncertainty

PROBLEMS

1.   You have just bought a raffle ticket for $1. There is one prize: $1000. Two thousand tickets were sold. What is the expected value of your purchase?

2. You have just bought an old desk for $10 at a garage sale. You think (you're 90 percent sure) that it is an antique worth $510. If it is not an antique (a 10 percent chance), you will have to spend another $20 to get it hauled away to the dump. What is the expected value of the desk?

3. The marketing department of a large company is trying to reach a decision about introducing a new product. They have compiled the following data:

Demand Level	Pr(Demand)	Anticipated Profit
High	.20	$300,000
Average	.50	$100,000
Low	.30	−$150,000

    a. Make a payoff table and find their best action (the actions are "introduce" or "don't introduce").

    b. Find the EVPI.

4. A floral shop sells an orchid an average of once every 5 days. They buy the orchids for $3 and sell them for $10. The demand for orchids follows a Poisson distribution. The flower is very fragile and will not keep for more than one day. Set up a payoff table for the number of orchids to be purchased each day.

    a. What are the actions?

    b. What are the events?

    c. Construct a payoff table and identify the proper action.

5. A commercial training institute is in the business of training computer programmers. They guarantee that they will find you a job within 48 hours of graduation or else refund your entire tuition. For every graduate they successfully place, they make a profit of $100. When the tuition is refunded, they lose $75. From their experience and by watching the want ads, they have determined that the demand for programmers during the 48-hour postgraduation time is a variable. They have made the following table:

Demand for Programmers	% of Time Demand Has Existed
15	10
16	30
17	40
18	15
19	5
	100

a. Set up a payoff table and determine the size of class this institute should have.

b. Find the EVPI.

6. The research and development department of a company is considering developing a new power source. From what is known of the theory right now, they think there is a 40 percent chance they could make it work. If it does work, there will be an estimated profit of $1 million for the company. If it doesn't work, they will have lost the estimated $200,000 it would cost to do the research. What should they do? Are there outside considerations that would have to be realistically taken into account?

7. You have just inherited a downtown building from your Uncle Seymour. The building is in an urban renewal area and will be razed in one year. You are looking at alternative ventures for making a profit on the building during the coming year. You would like to retain its present usage but the vice squad has already contacted you and says the girls must go. You can sell it right now for $5000 more than you will get from the urban renewal authorities. If the demand for storage space is high, you could realize $10,000 profit from renting the space. If the demand for storage space is average, your profits from renting would be $6000. And if the demand is low, your profits would be about $1000. After investigating the market you think the chance of high demand is .3, average demand is .5, and low demand is .2.

a. Set up a payoff table and determine the best action.

b. A consulting firm will predict the demand for storage space for a fee of $1000. Should you hire them? Why or why not?

8. A construction company is using four large diesel engines. They are concerned about the number of replacement engines they should store at the construction site. The site is relatively isolated, so that if an engine breaks down there is no chance of getting it repaired before the job is over. The engines ordered now cost $4000 (including freight and storage). But if an engine has to be ordered later and flown in, it will cost $6000 (because of the higher freight charges). An unused engine can be sold for $2000 at the completion of the job. Experience indicates a 25 percent chance that an engine will have a major breakdown. The breakdowns of engines are independent. Once an engine is replaced, it is virtually certain that it will not need to be replaced again.

  a. Set up a complete payoff table. (*Hint:* Compare this situation to the required characteristics of the binomial.)

  b. Find the maximum value of a guarantee of the number of breakdowns.

SELECTED REFERENCES

Spurr, William A. and Charles P. Bonini. *Statistical Analysis for Business Decisions.* Revised edition. Homewood, Ill.: Irwin, 1973. Especially Chapters 7, 8.

Tanur, Judith M., Frederick Mosteller, William H. Kruskal, Richard F. Link, Richard S. Pieters, and Gerald R. Rising. *Statistics: A Guide to the Unknown.* San Francisco: Holden-Day, 1972.

PART III

HOW TO ANALYZE STATISTICAL DATA

CHAPTER 8
SAMPLING AND SAMPLING DISTRIBUTIONS

OBJECTIVES: After studying this chapter and working the problems, you should be able to:

Explain why sampling is done and what is meant by statistical inference.

Discuss the primary methods of sampling and the advantages and disadvantages of each.

Define the concept of the central limit theorem.

Differentiate between a distribution of individual values and the distribution of the means of samples of a given size.

Define standard error.

State when the finite correction factor should be used.

THE CONCEPT OF sampling is used frequently in our daily lives:

We look out the window at a small part of the sky to "see how the weather is."

We take a small bite of a new dish to see if we like it.

After talking to a new acquaintance for a few minutes, we make a decision as to "what kind of a person she is."

Previews of new movies or television programs are shown to interest people in the whole film or show.

In selecting a mate we take a sample of what the person is like in different situations (the new generation seems to take a larger sample than the older generations).

Before introducing a new product, a company may test it on a relatively small group of people.

Rivers and lakes are sampled to check for pollution.

Sampling, briefly, is the examination of a part of a population. The purpose of sampling, generally, is to make a *statistical inference*. We examine a sample and from that examination we make judgments, or generalize, or make inferences about the entire population. In each of the preceding examples of sampling, an inference would be made about the population based on the analysis of a small sample.

Often it is mandatory that we take a sample rather than look at the entire population. It would be economically infeasible, for instance, to hire people to interview personally every consumer in the country about a new product. It would also be foolish for a manufacturer to test every light bulb he makes to see how long it will burn . . . he would never have any left to sell. What is surprising to many would-be analysts is the fact that in many cases a sample is *more* accurate than an enumeration of the entire population! A small, well-trained group of interviewers can usually get more accurate information from a sample than can a large group of untrained interviewers trying to talk to everybody. If we were trying to get information about a bond issue in a city, an attempt to talk to all the voters would lead to uncertainty about who will vote and who won't. Moreover, the time required would mean that some voters would have left town during the interviews while other potential voters had moved in (a localized population of people is a dynamic, fluid thing).

Sampling techniques are used in market research, quality control, and auditing, to name a few broad areas. Recently airlines and railroads have begun utilizing sampling to determine how much they owe each other for passengers and freight that travel on more than one company's equipment and routes. The accuracy of this sampling technique has proved to be sufficient, and the cost savings over enumerating each ticket and waybill are substantial.

We will formalize the techniques of making inferences in subsequent chapters. This chapter will serve, primarily, as a foundation for that later work. Understanding the concept of sampling and how to work with sampling statistics is imperative if we are to make logical statistical inferences.

WAYS OF TAKING A SAMPLE

A sample can be taken in myriad different ways. The major concern, though, is to take a sample in such a way that an analysis can be made of the accuracy of that sample. Sampling is a highly specialized field that can require a great deal of study involving higher mathematics. We will take a brief look at three general ways of taking a sample — random sampling, stratified sampling, and cluster sampling.

RANDOM SAMPLING

A *random sample* is merely a sample taken in such a way that each element in the population has an equal chance of being sampled. When raffle tickets are put into a large drum and thoroughly mixed, the drawing of the winning ticket is a random sample. When a part is drawn from a thoroughly mixed bin of such parts, it is a random sample.

A *systematic sample* is a special case of random sampling if the population is "thoroughly mixed." A systematic sample can be taken by randomly choosing a starting point and then taking every kth element after that. Dealing a deck of cards to four players, where a player gets every fourth card, is a systematic sample. In auditing, it is quite common for the auditor to take, say, every tenth invoice to get a 10 percent sample. In the 1970 census every twentieth person was asked a list of detailed questions. This was a systematic sample of 5 percent of the population. A systematic sample is more convenient and less costly than a completely random sample.

STRATIFIED SAMPLING

When a population is composed of distinct, identifiable groups, a stratified sample can be the most efficient method of sampling that population. A stratified sample selects a random sample of elements from *each* of the different groups or strata. The proportion of each stratum to be sampled depends on the variability of that stratum. The sampling method for the railroad waybills utilized stratified sampling: the waybills were stratified by the amount of each waybill, and different sampling procedures were used in each stratum.*

In industrial marketing, the potential customers for a new type of equipment may be stratified by "can use more than one of this equipment" and "can use only one of this equipment." A random sample of each of these strata would be taken. A consumer research sample is often stratified by income, region, family size, geographical region, or multiples of these.

The size of a sample depends on the variability of the items being sampled. Stratified sampling allows the analyst to increase the proportion sampled in strata with high variability and to decrease the total sampling cost by taking a small proportion of strata with small variability. At the same time, an adequate representation of all pertinent strata is guaranteed to appear in the overall sample.

CLUSTER SAMPLING

There are many variations of cluster sampling, but basically it amounts to randomly selecting *clusters* (elements in close proximity of each other) and then sampling elements in those clusters. The technique is used frequently in surveys of people requiring personal interviews. The interviewers can cut their travel time substantially by talking to a cluster of people in a small geographical territory. This reduction of travel time usually more than offsets the increased sample size this technique requires.

Although much more sophisticated sampling techniques are available, we will generally be referring to simple random samples in future discussions of a sample. You should be aware that the randomness of a sample is essential if you want to make reasonable inferences from the results. If there is no randomness involved in drawing your sample, then you have lost contact with the probabilities that are associated with making an accurate inference.

---

* For an easy to read summary of this project see: John Neter, "How Accountants Save Money by Sampling," in *Statistics: A Guide to the Unknown*, edited by Judith M. Tanur, et al. (San Francisco: Holden-Day, 1972).

THE CENTRAL LIMIT THEOREM*

In Chapter 6 we looked at probabilities associated with the normal distribution and concerned ourselves with how the individual elements of a population are distributed. This study was useful because many things seem to be normally distributed by nature. We now want to look at how *means of samples* are distributed. The *central limit theorem* gives us a great deal of help in determining probabilities associated with means of samples.

The central limit theorem is an amazing mathematical discovery. In general, it tells us that if we continue to take samples of a certain size n from an infinite population with mean μ and standard deviation σ, then if n is sufficiently large,

    1. The $\bar{x}$'s or means of these samples will tend to have a normal distribution.

    2. The mean of all the $\bar{x}$'s will be μ.

    3. The standard deviation of all the $\bar{x}$'s will be $\sigma_{\bar{x}}$, where $\sigma_{\bar{x}} = \sigma/\sqrt{n}$.

Perhaps we should examine how the central limit theorem works to clarify why it is so important for us in analytical work. First of all, the phrase "if n is sufficiently large" needs a little definition. Most statisticians use the general rule of thumb that n is sufficiently large for the conditions to hold if it is equal to or larger than 30 ($n \geqslant 30$).

Suppose that we are filling sacks of flour with automated machinery. The weight of the sacks is known to have a mean μ of 100 pounds and a standard deviation σ of 4 pounds. The sacks are shipped out in pallets of 64 sacks. What percentage of the pallets will have an average weight per sack of more than 101 pounds? (Perhaps the freight rate goes up if they reach such a weight.) The central limit theorem tells us that the averages ($\bar{x}$'s) of samples of size 64 will have approximately a normal distribution, and we can look at a pallet of 64 sacks as a sample of 64. It further tells us that the mean of the average weights of all pallets of 64 sacks will be μ, or 100 pounds. Finally, we

---

* You should review the discussion of the normal distribution in Chapter 6 before starting this section.

are told by the central limit theorem that the standard deviation of
the means of samples of size 64 will be

$$\sigma_{\bar{x}} = \frac{\sigma}{\sqrt{n}}$$

$$= \frac{4}{\sqrt{64}}$$

$$= .5$$

Putting all this together, we can find the answer to our question.
First we find the z value for means by

SO&B 8-1: $z = \dfrac{x - \mu}{\sigma_{\bar{x}}}$

Notice that this equation is similar to the formula for z values from
Chapter 6 except that we are now working with means ($\bar{x}$'s) of samples of a certain size rather than the raw data (x's) we used before.
This is the crux of the whole discussion, so it bears repeating. In
working with the normal distribution in Chapter 6, we were concerned about the distribution of the *individual* weights. Now we are
concerned about the distribution of the *means* of a certain sample
size. Continuing,

$$z = \frac{\bar{x} - \mu}{\sigma_{\bar{x}}}$$

$$= \frac{101 - 100}{.5}$$

$$= 2$$

From the tables for the normal distribution we find that .4772 of
the distribution will be between the mean and two ($z = 2$) standard
deviations above the mean. Therefore $.5000 - .4772 = .0228$ of the
distribution will be more than two standard deviations above the
mean — or .0228 of the pallets of 64 will have an average sack weight
of more than 101 pounds. This result is summarized in Figure 8-1.

Since the standard deviation of the means is called the *standard error*, we will use that term henceforth. Often we do not have the luxury
of knowing the population standard deviation σ. In these cases, we

usually estimate σ by the standard deviation S of an adequate sample size (again, $n \geq 30$).

Another important point must be made about the preceding example. We did not know how the original individual sack weights were distributed and, even more importantly, we really don't care! The central limit theorem tells us that regardless of the distribution of the individual items, the means will tend to be normally distributed as n gets large.

FIGURE 8-1.   Normal Distribution
($\mu = 100$, $\sigma_{\bar{x}} = .5$)

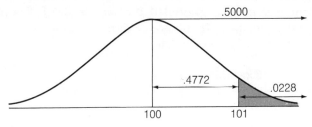

The fact that the standard error $\sigma_{\bar{x}}$ is equal to

$$\sigma_{\bar{x}} = \frac{\sigma}{\sqrt{n}}$$

can be proved mathematically for an infinite population. However, we can see intuitively that the means of samples of size n will not vary as much as the individual elements themselves because a sample of size 64, for instance, may include unusually high or low values of the individual items. These extreme values will, however, be "averaged out" or diluted by the other values in the sample. In other words, the larger the sample size, the less the means of that sample will vary from each other. At the extreme end, if we had a population of 1000 items and took samples of size 1000, then every time we took a sample we would obtain the entire population and each sample mean would be exactly the same. This would lead to a standard error of zero because there would be absolutely no variation in the sample means. At this extreme end we have introduced a finite sample, which leads us to the finite correction factor.

THE FINITE CORRECTION FACTOR

The size of the sample drawn has an additional effect on the standard error. If we are drawing samples of size n from a finite population of size N, then the standard error is

SO&B 8-2:
$$\sigma_{\bar{x}} = \frac{\sigma}{\sqrt{n}} \sqrt{\frac{N-n}{N-1}}$$

The term $\sqrt{(N-n)/(N-1)}$ is called the *finite correction factor*. If n is very small relative to N, the finite correction is not going to have a significant effect on $\sigma_{\bar{x}}$. As a rule of thumb, then, the finite correction factor is generally ignored if the sample size is less than 5 percent of the population. If the population is infinite in size, the finite correction factor is never needed.

> **Example 1:** In the discussion of the normal distribution in Chapter 6, we worked an example concerning bricks that weighed an average of 5.0 pounds. These weights had a standard deviation of .20 and were normally distributed. In that example we found that the probability of a customer randomly selecting a brick which weighed between 5.0 and 5.2 pounds was .3413 [$\Pr(5.0 \leqslant x \leqslant 5.2) = .3413$].
>
> Now suppose that a customer is planning on building a brick outhouse that requires 100 bricks. If the building weighs too much, it will sink into the ground. If it doesn't weigh enough, it will blow over. The customer therefore

FIGURE 8-2.   Normal Curve ($\mu = 5$, $\sigma = .20$)

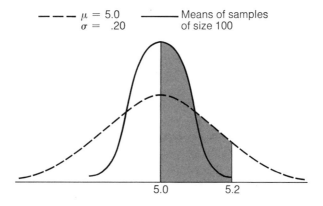

wants the 100 bricks to have an average weight of between 5.0 and 5.20 pounds, which is the same weight the customer wanted one brick to weigh in Chapter 6. What is the probability that the customer can randomly pick 100 bricks whose average weight fits his requirement? First of all, we know that the average weight of samples of size 100 will follow a normal distribution. Using SO&B 8-1, we find that

$$z = \frac{\bar{x} - \mu}{\sigma_{\bar{x}}}$$

$$= \frac{5.20 - 5.0}{.20/\sqrt{100}}$$

$$= \frac{.20}{.02}$$

$$= 10$$

A z value of 10 does not even appear in the table—which means, for all practical purposes, that .5000 of the average weights of samples of size 100 will be between 5.0 and 5.20 pounds. The comparison between the individual weight distribution and the average weight distribution is shown in Figure 8-2.

Conversely, suppose the customer had wanted 100 bricks whose average weight was between 5.0 and 5.04 pounds. Now, using SO&B 8-1, we find that

$$z = \frac{5.04 - 5.0}{.20/\sqrt{100}}$$

$$= \frac{.04}{.02}$$

$$= 2$$

From Table A-3 in the appendix we find that

$$\Pr(5.0 \leq \bar{x} \leq 5.04) = .4772$$

However, if upon finishing the outhouse he finds that he needs one more brick, what is the probability that he can

come back and randomly select one brick that weighs between 5.0 and 5.04 pounds? Using SO&B 6-3 we find that

$$z = \frac{x - \mu}{\sigma}$$

$$= \frac{5.04 - 5.0}{.20}$$

$$= \frac{.04}{.20}$$

$$= .20$$

From Table A-3 we find that

$$\Pr(5.0 \leqslant x \leqslant 5.04) = .0793$$

The comparison of these values is shown in Figure 8-3.

FIGURE 8-3.   Normal Curve ($\mu = 5$, $\sigma = .20$)

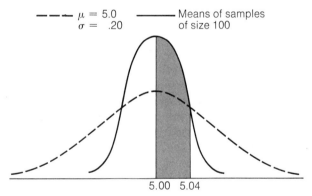

To demonstrate the application of the finite correction
factor, let's revise the preceding problem slightly. Instead
of having a virtually infinite production of bricks, let's consider what happens if we have only a stock of 300 bricks.
Now a sample of 100 bricks is more than 5 percent of the
population of 300 bricks. Hence we want to apply the
finite correction factor. Using SO&B 8-2,

$$\sigma_{\bar{x}} = \frac{\sigma}{\sqrt{n}} \sqrt{\frac{N-n}{N-1}}$$

$$= \frac{.20}{\sqrt{100}} \sqrt{\frac{300-100}{300-1}}$$

$$= .02 \sqrt{\frac{200}{299}}$$

$$= .02 \sqrt{.669}$$

$$= .02 \cdot .818$$

$$= .016$$

To find $\Pr(5.00 \leq \bar{x} \leq 5.04)$, we apply SO&B 8-1:

$$z = \frac{5.04 - 5.00}{.016}$$

$$= \frac{.04}{.016}$$

$$= 2.50$$

From Table A-3 we then find $\Pr(5.00 \leq \bar{x} \leq 5.04) = .4938$. Notice that this result is considerably different from the previous situation of an infinite population, where we found that

$$\Pr(5.00 \leq \bar{x} \leq 5.04) = .4772$$

AUTOPSY

In subsequent work we will formalize some concepts of statistical inference. All of the preceding examples are based on the assumption that a logical sampling technique is used. Statistical inference can be thought of as looking at the sky through windows in a house to determine the weather conditions. If you look through one small window and see that it is raining, you can make a reasonably certain inference that it is raining all over outside. If you look through one small window and see perfectly clear sky — absolutely no clouds — you would be willing to infer that it is a sunny day. But if you look out the window and see some clouds, you will probably want to look out other windows (take a larger sample) before you make any inference about the weather. The more variable the population, the larger the sample that is required in making inferences. Put another way, the more variable the sample, the less confident we are about any inferences to be made.

Remember that in this chapter we have begun to work with means of samples of a certain size instead of individual elements. Much of your ability to analyze statistical data intelligently will depend on your ability to apply the normal distribution. And the application of the normal distribution is made much more feasible by the central limit theorem.

SUMMARY

Common Sampling Techniques

1. Random—all elements have an equal chance of being chosen.

2. Systematic—a special case of random sampling in which every kth element is chosen.

3. Stratified—the sample is drawn from each of the identifiable strata in the population.

4. Cluster—the sample is taken relative to close groups of elements.

The central limit theorem tells us that means of samples of size n tend to have a normal distribution with mean μ and standard error $\sigma_{\bar{x}}$, where $\sigma_{\bar{x}} = \sigma/\sqrt{n}$ if the population being sampled is infinite in size.

The finite correction factor is used to "adjust" the standard error when the sample is more than 5 percent of the population:

$$\sigma_{\bar{x}} = \frac{\sigma}{\sqrt{n}} \sqrt{\frac{N-n}{N-1}}$$

TABLE 8-1. Statistical Notation

Item	For a Population	For a Sample
Size	N	n
Mean	μ	$\bar{x}$
Standard deviation	σ	S
Standard error	$\sigma_{\bar{x}}$	$S_{\bar{x}}$

PROBLEMS

1. How can opinion surveys and market research surveys of relatively few people result in statements or predictions about large numbers of people?

2. A consumer advocate group wants to look at new but unsold cars on a large dealer's lot to estimate the percentage of cars that have serious defects. What kind of sample should they take—random, stratified, or cluster? Why?

3. A lumber dealer who sells both retail and wholesale has a file of all his sales invoices to customers. He wants to know the average dollar amount of the invoices. Should he take a random, stratified, or cluster sample? Why?

4. A market research team wants to take a sample of a large metropolitan area to determine how people react to a new product. This survey requires a personal interview. Would you suggest a random, stratified, or cluster sample? Why?

5. A certain process on an assembly line is known to take an average of .35 minute with a standard deviation of .14 minute. A sample of 49 such processes is observed.

> a. What is the probability that the average time of the sample will be between .35 and .40 minute?
>
> b. What is the probability that the average time of the sample will be less than .33 minute?
>
> c. What is the probability that the average time of the sample will be less than .33 minute or greater than .40 minute?

6. A customer buys 100 hydraulic hoses from a manufacturer who claims that the hoses will last an average of 1000 hours with a standard deviation of 50 hours. The customer keeps accurate records on the 100 hoses and finds that the average life is 975 hours. Do you believe the manufacturer's claim?

7. Define *standard error* and explain the difference between a standard error and a standard deviation.

8. Refer to Problem 15 in Chapter 6. Answer parts (*a*) and (*b*) relative to the average value of a sample of 40 hoses. Compare these answers with the answers to the original problem involving individual hoses by drawing appropriate sketches.

9. A company is marketing lawn fertilizer in 80-pound bags. Records show that the bags have an average weight of 80 pounds with a standard deviation of 2.4 pounds. The bags are stored on pallets of 64 bags. The pallets are moved and loaded with forklifts. A cheap forklift can be obtained, but the maximum weight it can handle is 5130 pounds. Should they buy the forklift? Why? (*Hint:* Convert the total weight of 5130 pounds to an average bag weight for a sample of 64.)

10. The lawn fertilizer company in Problem 9 above has 512 bags left in stock.

> *a.* What proportion of the pallets of 64 bags would you expect to have an average bag weight of between 80 and 80.5 pounds?
>
> *b.* What proportion of the pallets of 64 bags would you expect to have an average bag weight of 79.5 to 80.5 pounds?

11. Suppose the amount of detergent used daily by a hospital is normally distributed with a mean of 80 pounds and a variance of 9 pounds. What proportion of the days will they use more than 81 pounds? A random sample of 64 of the daily use records is taken. What is the probability that the average daily weight used will be more than 81 pounds? Draw a sketch comparing these two situations and the answers.

12. Write out the central limit theorem and explain its importance.

SELECTED REFERENCES

Bryant, Edward C. *Statistical Analysis*. Revised edition. New York: McGraw-Hill, 1966.

Cochran, William G. *Sampling Techniques*. Second edition. New York: Wiley, 1963.

Guenther, William C. *Concepts of Statistical Inference*. Second edition. New York: McGraw-Hill, 1973. Especially Chapter 3.

Miller, Irwin and John E. Freund. *Probability and Statistics for Engineers*. Englewood Cliffs, N.J.: Prentice-Hall, 1965. Especially Chapter 7.

Spurr, William A. and Charles P. Bonini. *Statistical Analysis for Business Decisions*. Revised edition. Homewood, Ill.: Irwin, 1973. Especially Chapters 9, 12.

Tanur, Judith M., Frederick Mosteller, William H. Kruskal, Richard F. Link, Richard S. Pieters, and Gerald R. Rising. *Statistics: A Guide to the Unknown.* San Francisco: Holden-Day, 1972.

Wallis, W. A. and H. V. Roberts. *The Nature of Statistics.* New York: Free Press, 1962. Especially Chapter 5.

Yamane, Taro. *Statistics: An Introductory Analysis.* Third edition. New York: Harper & Row, 1973. Especially Chapter 7.

CHAPTER 9
ESTIMATION

OBJECTIVES: After studying this chapter and working the problems, you should be able to:

Define what is meant by a point estimate.

Explain how and why an interval estimate yields more information than a point estimate.

Demonstrate how to use a confidence interval in making an interval estimate.

Calculate required sample size at a stipulated accuracy.

ONE OF THE purposes of statistics is to make estimations about population parameters. We seldom know the value of population parameters, but we try to estimate them by appropriate sample statistics. Remember: A parameter is a characteristic of a population and a statistic is a characteristic of a sample.

Estimation involves sampling, which we discussed briefly in Chapter 8. Normally we accept the mean of a sample, $\bar{x}$, as an estimate of the population mean μ. If we are concerned about a proportion—say the proportion of consumers who prefer our product in a white package or the proportion of voters who favor a certain political candidate—we generally accept the proportion of such people in a sample, $\bar{p}$, as an estimate of the proportion of such people in the population, P. Henceforth we will use the notation of $\bar{p}$ for the proportion in a sample and P for the proportion in a population.

Estimates can be divided into two general groups: point estimates and interval estimates. Generally we are exposed to point estimates (a single value), such as "the average purchase amount is about $5" or "about 40 percent of our employees would prefer a four-day work week." The point, or single-value, estimate is lacking in information, though, since it does not give you any idea of how accurate you can expect this estimate to be.

To illustrate this concept, suppose you are trying to estimate Mable Metcher's age. You ask an expert on guessing ages to give you such an estimate and, after a thorough examination (well, not *too* thorough), he guesses her age to be 30. Then suppose you ask a drunk standing on the corner to guess her age. After a considerable effort to focus his eyes the drunk says, "I'd guessh her age to be about shirty." Both sources (samples) gave you the same point estimate, but is there a difference in the amount of confidence you would have in each of these two estimates? After the expert's estimate you might say to yourself, "Well, her age is probably between 29 and 31 years." But after the drunk's estimate you would probably think something like, "Well, her age is probably between 15 and 45 years."

The estimates of "29 to 31 years" and "15 to 45 years" are *interval estimates.* Interval estimates involve a group or interval of values. Note that there is more intuitive information contained in an interval estimate than in a point estimate. With an interval estimate you get a feel for how accurate the estimate may be or how much confidence you can have in the sampling procedure. The interval estimate is, in actuality, an extremely underused description. The point estimate, which we are normally exposed to, does not give us any idea of the possible magnitude of *sampling error*—the difference between the sample statistic and the population parameter. In fact, we rarely know the actual sampling error because we rarely know the true population parameter. If we knew the population parameter, we wouldn't be wasting our time estimating it with a sample statistic in the first place.

Learning to think and analyze in terms of interval estimates can be valuable to you in many ways. Suppose you are working with a new product. An engineer makes a formal estimate that 100,000 units of this new product can be sold. This point estimate of 100,000 units does not give you enough information for action. If your engineer is fairly certain of his estimate, further questioning may reveal that he expects between 90,000 and 110,000 sales. If he is relatively uncertain about his estimate, he might state that he expects between 50,000 and 150,000 sales. An interval estimate would tell you, the analyst, a great deal more than the original point estimate. It may lead you to decide how much further to pursue the project and how much more work needs to be done on estimating sales for this new product.

The error of a point estimate is normally an implied error. When a single figure is given, it implies that the figure is absolutely accurate. Common sense tells us, however, that we would be extremely surprised if an estimate based on a sample was *exactly* equal to the population parameter being estimated. How inexact is it, though? A common technique for handling this problem is confidence intervals.

CONFIDENCE INTERVALS

Suppose we were to begin with a point estimate gained from a sample and then expanded that to an interval estimate in such a way that we could attach a probability statement such as Pr(the interval estimate contains the population parameter) = a value. Then our end result (the probability statement) would be a *confidence interval.*

CONFIDENCE INTERVALS FOR QUANTITIES

How do you go about setting up a confidence interval? We now have at our disposal the tools to make a confidence interval by using a statistic that we can assume to be approximately normally distributed. We also know from the central limit theorem that means of samples of size greater than 30 tend to be normally distributed. If we wanted a confidence interval concerning a population mean μ, we could devise such a creation by using the sample mean $\bar{x}$ if the sample were sufficiently large. Here is the step-by-step procedure we would use.

Step 1:   Find $\bar{x}$ from the sample.

Step 2:   If σ is unknown, estimate it by finding S, the sample standard deviation.

Step 3:   Decide on the desired confidence level. The confidence level is the probability that the resulting confidence interval includes μ, the population mean. The confidence level is usually set at 99, 95, or 90 percent. This confidence level is denoted by $1 - \alpha$, where α is the probability that the confidence interval does *not* contain μ.

Step 4:   Assuming that the sample is larger than 30, find the z value in the normal tables that cuts off $\alpha/2$ proportion of the data. For a confidence level of 95 percent (that is, $1 - \alpha = .95$ and $\alpha = .05$), Figure 9-1 shows this process.

FIGURE 9-1. z Values for 95 Percent
Confidence Interval

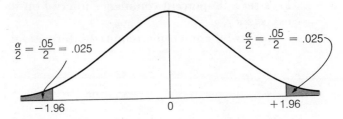

In the normal table we see that a z value of 1.96 cuts off
2.5 percent ($\alpha/2$) of the population in one tail of the normal
curve. Therefore there is a total of 95 percent of the population between -1.96 and $+1.96$ standard deviations. Other
z values for commonly used confidence levels are shown
in Table 9-1.

TABLE 9-1. z Values for
Common Confidence Levels

Confidence Level $(1 - \alpha)$	z Value
.99	2.57
.95	1.96
.90	1.65

Step 5: Using steps 1 to 4, calculate the confidence interval:

SO&B 9-1 $\Pr(\bar{x} - z\sigma_{\bar{x}} < \mu < \bar{x} + z\sigma_{\bar{x}}) = 1 - \alpha$

Example 1. We are a company that markets stereo units.
Most of the components of the units we assemble are
purchased from supplier companies. A new supplier has
submitted a bid to supply a component. One variable we
are particularly concerned about is the life expectancy of
each component, so we take a sample of 100 of the new
supplier's components and send them to our testing area
to determine the life expectancy. The testing area reports
that the sample of 100 had an average life expectancy of

495 hours with a standard deviation of 64 hours. The mean $\bar{x}$ of 495 hours is a point estimate of the life expectancy. Let's set a 95 percent confidence interval on the true life expectancy, μ.

We have the following required information:

$$\bar{x} = 495$$

$$n = 100$$

$$1 - \alpha = .95$$

Since we have a large sample, we will accept $S = 64$, the standard deviation of the sample, as a reliable estimate of σ, the standard deviation of the population. Plugging this information into SO&B 9-1, we have

$$\Pr(\bar{x} - z\sigma_x < \mu < \bar{x} + z\sigma_{\bar{x}}) = 1 - \alpha$$

$$\Pr\left(495 - 1.96 \cdot \frac{64}{\sqrt{100}} < \mu < 495 + 1.96 \cdot \frac{64}{\sqrt{100}}\right) = .95$$

$$\Pr(495 - 1.96 \cdot 6.4 < \mu < 495 + 1.96 \cdot 6.4) = .95$$

$$\Pr(495 - 12.5 < \mu < 495 + 12.5) = .95$$

$$\Pr(482.5 < \mu < 507.5) = .95$$

The technical translation of a 95 percent confidence interval is this: 95 percent of the time that we appropriately calculate a confidence interval in this fashion, the confidence interval will include the true population mean. In practical usage the analyst will sometimes slip and say that he's 95 percent confident the population mean is between 482.5 and 507.5 hours. As for me, I have yet to feel the earth stop moving on such an occasion, even though this is not a correct interpretation of a confidence interval.

Example 2:   Let's return to our brick company. Suppose we know that the standard deviation of the weights of the bricks is .20 pound, but we do not know the average weight. A sample of 50 bricks shows an average weight of 5.25 pounds. To construct a 90 percent confidence interval for the true weight, we would use SO&B 9-1:

$$\Pr(\bar{x} - z\sigma_{\bar{x}} < \mu < \bar{x} + z\sigma_{\bar{x}}) = 1 - \alpha$$

$$\Pr\left(5.25 - 1.65 \cdot \frac{.20}{\sqrt{50}} < \mu < 5.25 + 1.65 \cdot \frac{.20}{\sqrt{50}}\right) = .90$$

$$\Pr\left(5.25 - 1.65 \cdot \frac{.20}{7.07} < \mu < 5.25 + 1.65 \cdot \frac{.20}{7.07}\right) = .90$$

$$\Pr(5.25 - 1.65 \cdot .028 < \mu < 5.25 + 1.65 \cdot .028) = .90$$

$$\Pr(5.25 - .05 < \mu < 5.25 + .05) = .90$$

$$\Pr(5.20 < \mu < 5.30) = .90$$

CONFIDENCE INTERVALS FOR PROPORTIONS

We also need to put confidence intervals around population proportions. The only difference when we are working with proportions is that we can calculate the standard error for a sample proportion by the formula

$$\sigma_{\bar{p}} = \sqrt{\frac{\bar{p}\bar{q}}{n}} \qquad (\text{Remember: } \bar{q} = 1 - \bar{p})$$

which we used in Chapter 6. The concept is the same as the means, though. If the sample size is sufficiently large (recall Chapter 6), we can calculate a confidence interval for a population parameter by going through the following five steps.

Step 1: Find $\bar{p}$ from the sample.

Step 2: Calculate $\sigma_{\bar{p}} = \sqrt{\dfrac{\bar{p}\bar{q}}{n}}$

Step 3: Decide on the appropriate confidence level, $1 - \alpha$.

Step 4: Find the z value in the normal tables that cuts off $\alpha/2$ proportion of the data in one tail of the normal distribution.

Step 5: Using steps 1 to 4, calculate the confidence interval:

SO&B 9-2: $$\Pr(\bar{p} - z\sigma_{\bar{p}} < P < \bar{p} + z\sigma_{\bar{p}}) = 1 - \alpha$$

Example 3: A distributor has just bought a large inventory of hose fittings from a defunct company and would like to estimate the proportion of this purchase that is brass as opposed to aluminum or plastic. He looks at a random sample of 50 from this huge assortment and finds that 20 in the sample are brass. He wants to prepare a 99 percent confidence interval around the true proportion of brass fittings in the entire purchase. First we have found in our sample that

$$\bar{p} = \frac{20}{50}$$

$$= .40$$

We can then use SO&B 9-2 to construct the confidence interval:

$$\Pr(\bar{p} - z\sigma_{\bar{p}} < P < \bar{p} + z\sigma_{\bar{p}}) = 1 - \alpha$$

$$\Pr\left[.40 - 2.57 \cdot \sqrt{\frac{.40(.60)}{50}} < P < .40 + 2.57 \cdot \sqrt{\frac{.40(.60)}{50}}\right] = .99$$

$$\Pr(.40 - 2.57 \cdot .069 < P < .40 + 2.57 \cdot .069) = .99$$

$$\Pr(.40 - .18 < P < .40 + .18) = .99$$

$$\Pr(.22 < P < .58) = .99$$

Example 4: By this time Mabel Metcher's computerized dating business has become so successful that she has quit her job and opened a full-time business. Mable Metcher's Mate Matcher has now been shortened to The 4M Company. A dissatisfied customer who also happens to be a consumer advocate has become concerned about the proportion of 4M customers who are unhappy with the service. He talks to a random sample of 36 customers and find that 9 of them are dissatisfied. The consumer advocate then confronts Mabel with the information that in his sample the proportion of dissatisfied customers was $\bar{p} = \frac{9}{36} = .25$. Mabel replies that this is only a point estimate and he should get more information concerning the reliability of his estimate by constructing a confidence interval. The consumer advocate sits down in his store-front office and calculates:

$$\Pr(\bar{p} - z\sigma_{\bar{p}} < P < \bar{p} + z\sigma_{\bar{p}}) = 1 - \alpha$$

$$\Pr\left[.25 - 1.96 \cdot \sqrt{\frac{.25(.75)}{36}} < P < .25 + 1.96 \cdot \sqrt{\frac{.25(.75)}{36}}\right] = .95$$

$$\Pr(.25 - 1.96 \cdot \sqrt{.0052} < P < .25 + 1.96 \cdot \sqrt{.0052}) = .95$$

$$\Pr(.11 < P < .39) = .95$$

He gives this answer to Mabel and tells her that a point estimate of the number of days she'll spend in court is 60 . . . at a 100 percent confidence level!

DETERMINING SAMPLE SIZE

Thus far in our discussion we have just assumed a certain sample size. When we are beginning a problem of estimating some parameter, though, the question of "How large a sample do I need?" almost always arises. The size of the sample is an important factor because we want to take a sample that is large enough to satisfy our requirements for accuracy. However, the cost of taking a sample rises with the number of items we sample. Therefore we do not want to waste money by arbitrarily taking a very large sample.

DETERMINING SAMPLE SIZE FOR QUANTITIES

If the sample statistic is normally distributed (remember that we have the central limit theorem at our disposal), we can estimate the required sample size by using the following four steps.

Step 1:   Determine the accuracy you require. If you are trying to estimate the average purchase amount of a large group of customers, you may want the estimate to be within 50 cents (± 50 cents) of the true population average. If you are trying to estimate the weight of bricks, you may want the estimate to be within .1 pound ($\pm .1$ pound) of the true population mean. If you are trying to estimate the life expectancy of batteries, you may want to estimate within 10 hours (± 10 hours) of the true average life expectancy. The accuracy that you determine will be denoted by E. In these examples $E = .50, E = .1$, and $E = 10$. The magnitude of E will depend on how crucial an accurate estimate is. The smaller that E is required to be, the larger the sample size required. The larger that E may be, the smaller the sample size required. The value of E is a subjective decision on the part of the manager or analyst.

Step 2:   Decide on a confidence coefficient. This is usually 99, 95, or 90 percent. In many business applications 90 percent is a reasonable value. This value is, like E in step 1, determined subjectively and depends on the critical nature of the estimate. The higher the confidence level desired, the larger the sample size required. The corresponding z value is then found, just as in the procedure for confidence intervals.

Step 3:   If the population standard deviation σ is unknown, estimate it in some way. You could do this by calculating S from a large sample or perhaps merely by past experience with similar analyses. If the estimate is shown later, by the results of the sample, to be inaccurate, the estimate of σ can then be revised and the required sample size recalculated.

Step 4:   Put steps 1 to 3 into the formula

SO&B 9-3:
$$n = \frac{z^2\sigma^2}{E^2}$$

to find the minimum required sample size n. The formula $n = z^2\sigma^2/E^2$ can be derived algebraically from the formula $z = (\bar{x} - \mu)/\sigma_{\bar{x}}$, where the value of E is $\bar{x} - \mu$. Notice here that E, the accuracy required, is actually the difference between the sample mean $\bar{x}$ and the population mean μ. If you are interested in the derivation of SO&B 9-3, see Problem 10 at the end of the chapter.

Example 5: Suppose once again that we are marketing stereo units and that we purchase from supplier companies the components we assemble. A new supplier submits a bid to supply a component. Now we want to take a sample of his product to estimate the life expectancy. We determine that we would like to estimate the life expectancy within ±10 hours of the true life expectancy with a 95 percent confidence coefficient. Similar products have shown a standard deviation σ of 64 hours, and we are willing to accept this as an original estimate of σ for this component. From this information we know that the z value for a 95 percent confidence coefficient is 1.96, $\sigma = 64$, and the required accuracy is $E = 10$. Using SO&B 9-3 we would calculate

$$n = \frac{z^2\sigma^2}{E^2}$$
$$= \frac{(1.96)^2(64)^2}{(10)^2}$$
$$= 157.35$$

We would need, then, a sample of 157 units from the new supplier to make an estimate that would fit our accuracy requirements.

Example 6: If we wanted to sample bricks to estimate the average weight and we knew that $\sigma = .20$, we could determine the sample size needed by first determining how

$$\frac{(1.96)^2 (60)^2}{5\%^2} = 558$$
$$= 392 \text{ nesk.}$$

accurate the estimate must be. Suppose we decide that we want the sample mean to be within $\pm.05$ pound of the true population average (that is, $E = .05$) with a 90 percent confidence level. To find the sample size we would calculate

$$n = \frac{(1.65)^2(.20)^2}{(.05)^2}$$

$$= 43.56$$

We would need a sample of at least 44 bricks to meet our stipulated accuracy requirements.

DETERMINING SAMPLE SIZE FOR PROPORTIONS

If we are concerned about estimating a proportion, the procedure for determining the appropriate sample size is similar to the one we used for quantities. In determining the appropriate sample size for proportions, we use the same steps as those for determining sample sizes for quantities with the exception of step 3, which involved estimating σ. In place of the calculated or estimated σ, we use $P \cdot Q$, which is the variance for proportions. The formula to be used in finding an appropriate sample size for estimating a proportion is

SO&B 9-4: $n = \dfrac{z^2 PQ}{E^2}$

This formula can be derived from the formula

$$z = \frac{\bar{p} - P}{PQ/n}$$

If you are interested in the derivation, see Problem 11 at the end of the chapter. The P in this formula is our estimate of what we suspect the population proportion might be. If we have absolutely no idea of what the proportion might be, we use $P = .50$ because this value will give us the maximum sample size required. In that way, we'll know we have at least enough sample size. The Q of course, is $1 - P$.

> **Example 7:**   We worked previously with a distributor who had just bought a huge inventory of hose fittings from a defunct company and wanted to estimate the proportion

of fittings that were brass. Now suppose that he wants this estimate to be within ±5 percentage points (±.05) of the true proportion (that is, $E = .05$) and wishes to be 99 percent confident in his answer. Suppose also that he believes, either from a visual inspection or from the seller, that the proportion of brass fittings is about .40. This .40 will be our estimate of P, and using SO&B 9-4 to estimate the required sample size, we have

$$n = \frac{z^2 PQ}{E^2}$$

$$= \frac{(2.57)^2 \cdot .40 \cdot .60}{(.05)^2}$$

$$= 634.07$$

The distributor will need a sample of 634 fittings to estimate with the stipulated accuracy the proportion that are brass.

Example 8: The consumer advocate who is harassing The 4M Company realizes that he must have statistically valid information concerning the proportion of dissatisfied customers if he is to succeed in bringing action against Mabel. He is concerned about how large a sample he needs to estimate the true proportion within ±10 percentage points of the true percentage (that is, $E = .10$) at a 95 percent confidence level. At this point he has absolutely no idea of what the true proportion of dissatisfied customers might be, so he lets $P = .50$ and uses SO&B 9-4 to find that

$$n = \frac{z^2 PQ}{E^2}$$

$$= \frac{(1.96)^2 \cdot .50 \cdot .50}{(.10)^2}$$

$$= 96.04$$

To obtain the stipulated accuracy, he needs to take a random sample of 96 customers of The 4M Company to estimate the proportion of dissatisfied customers.

AUTOPSY

In this chapter we have seen once again that the field of statistics is not as cut and dried as some people would have you believe. The calculation of the formulas may be straightforward, but we have dealt with some subjective decisions, such as choosing a confidence level and deciding on a desired accuracy level. These are the situations the would-be analyst must face. One item you should watch in doing the calculations in this chapter is the decimal points: many students forget to include the decimal points when they are working with confidence intervals or determining sample sizes involving proportions.

SUMMARY

Confidence Intervals for Quantities

SO&B 9-1: $\Pr(\bar{x} - z\sigma_{\bar{x}} < \mu < \bar{x} + z\sigma_{\bar{x}}) = 1 - \alpha$

Confidence Intervals for Proportions

SO&B 9-2: $\Pr(\bar{p} - z\sigma_{\bar{p}} < P < \bar{p} + z\sigma_{\bar{p}}) = 1 - \alpha$

where $\sigma_{\bar{p}} = \sqrt{\bar{p}\bar{q}/n}$

Sample Size for Quantities

SO&B 9-3: $n = \dfrac{z^2\sigma^2}{E^2}$

Sample Size for Proportions

SO&B 9-4: $n = \dfrac{z^2 PQ}{E^2}$

Some Confidence Coefficients and Corresponding z values

$1 - \alpha$	z
.99	2.57
.95	1.96
.90	1.65

TABLE 9-2.   Statistical Notation

Item	For a Sample	For a Population
Proportion	$\bar{p}$	P
Standard error of proportion	$S_{\bar{p}}$	$\sigma_{\bar{p}}$

PROBLEMS

1.   An electronic device is being tested for life expectancy. The standard deviation of the life expectancy is known to be 64 hours. A sample of 100 shows an average life expectancy of 495 hours.

   a.   What is the point estimate of the average life expectancy?

   b.   Construct a 99 percent confidence interval for μ, the true average life expectancy.

   c.   Does part (*b*) contain more information than part (*a*)?

   d.   Construct a 90 percent confidence interval for μ, the true average life expectancy.

   e.   Compare the intervals found in parts (*b*) and (*d*) above. Is it logical that the interval in (*b*) is larger than that found in (*d*)? Why?

2.   George Gouge, president of the Nocturnal Aviation Land Company, is interested in finding out the average length of time it takes to close a sale with his special sales techniques. He keeps track of the time it takes his salesmen to close a random sample of 64 sales. The average time in the sample is 73 minutes with a standard deviation of 20 minutes.

   a.   Set a 95 percent confidence interval for the true time it takes to close a sale using the special techniques.

   b.   Suppose these results were from a sample of 150 sales instead of 64. Now do part (*a*) again.

   c.   Compare the answers to parts (*a*) and (*b*). Is it logical that the interval in (*a*) is larger than that found in (*b*)? Why?

3.   A sample of 100 full-time workers in an urban area shows an average hourly wage of $4.94. The standard deviation of the hourly

wages is $1.10. Construct a 99 percent confidence interval for the mean.

4.   A recent survey of 100 citizens advocating "law and order" legislation shows that 70 percent of them sometimes exceed the speed limit while driving. Construct a 95 percent confidence interval for this proportion.

5.   Assume that after doing the survey in Problem 4, it is decided that the population proportion should be estimated within ±1 percentage point (±.01) with a 95 percent confidence level.

   a.   How large a sample would be required?

   b.   How large a sample would be required if we only needed to measure the proportion within ±5 percentage points?

6.   A random sample of 100 residents contains 40 who believe that the public school system is at least adequate. Set a 95 percent confidence interval on the proportion of the population who believe that the system is at least adequate.

7.   A vending machine company is interested in determining the number of cups of coffee required daily by each office of a large office building. From past experience they know that the standard deviation of the demand per day is 5. They would like to make their estimate within a range of ±1 cup at a 95 percent confidence level. How large a sample should they take in trying to make their estimate of daily demand per office?

8.   A state welfare agency wants to estimate the average amount of time prospective clients are kept waiting in the office before their application interview. They would like to estimate this time within $1\frac{1}{2}$ minutes of the true average and be 90 percent confident in their result. It is estimated that the standard deviation of this waiting time is 15 minutes. How large a sample do they need to meet the stipulated accuracy requirements?

9.   A medical research laboratory has developed a new treatment for a certain malady. They would like to take an actual sample to determine the proportion of people with the malady that will respond favorably to the treatment. They wish to be 99 percent confident in their result and would like their sample estimate to be within $\frac{1}{2}$ percent of the true proportion. How large a sample do they need if

initial experimentation leads them to believe that about 70 percent
will respond favorably?

10. Derive SO&B 9-3 from

$$z = \frac{\bar{x} - \mu}{\sigma_{\bar{x}}}$$

11. Derive SO&B 9-4 from

$$z = \frac{\bar{p} - P}{\sqrt{PQ/n}}$$

12. Larry Lisp, the market research director for a manufacturer of
sea shell charms, has been made aware that the new plastic being used
will suddenly deteriorate so that the sea shell charms chip. He has
received a great number of complaints. Selecting a random sample
of 50 of these complaints, Larry goes out to conduct personal interviews and determine how many of these customers with defective
merchandise will never buy his product again. After being slugged
six times (he had trouble phrasing the questions), Larry found 20
customers who swore they would never buy his product again.

   a. Set a 95 percent confidence interval on the appropriate parameter.

   b. If Larry wanted to be 95 percent sure that he was
   within ±10 percentage points of the true parameter, how
   large a sample should he take?

13. A manufacturer of children's toys would like to know the average height of four-year-old boys. He knows from past studies that
the variance of these heights is about 16 inches. He must estimate
the average height within 1 inch and be 95 percent confident in his
results. What is the appropriate sample size?

14. Greta Goldheart is a well-educated barmaid who tutors failing
students for extra income. Greta knows how much money she needs
in total, so she wants to translate this figure to the number of students she has to tutor. This number, of course, depends on the average amount she gets from each student. Looking over her records,
which she keeps for income tax purposes, she finds that the last 50
students yielded an average of $4.23 with a variance of $1.21. There

are 500 failing students in the neighborhood and she never tutors the same one twice.

    a. Set a 95 percent confidence interval for the average amount of yield. (*Hint:* Don't forget the finite correction factor.)

    b. What would be the confidence coefficient if Greta wanted to set her confidence limits at \$4.23 ± \$0.10?

SELECTED REFERENCES

Bryant, Edward C. *Statistical Analysis.* Revised edition. New York: McGraw-Hill, 1966.

Guenther, William C. *Concepts of Statistical Inference.* Second edition. New York: McGraw-Hill, 1973. Especially Chapter 5.

Miller, Irwin and John E. Freund. *Probability and Statistics for Engineers.* Englewood Cliffs, N.J.: Prentice-Hall, 1965. Especially Chapter 8.

Tanur, Judith M., Frederick Mosteller, William H. Kruskal, Richard F. Link, Richard S. Pieters, and Gerald R. Rising. *Statistics: A Guide to the Unknown.* San Francisco: Holden-Day, 1972.

Wallis, W. A. and H. V. Roberts. *The Nature of Statistics.* New York: Free Press, 1962. Especially Chapter 8.

Yamane, Taro. *Statistics: An Introductory Analysis.* Third edition. New York: Harper & Row, 1973. Especially Chapter 8.

CHAPTER 10
TESTS OF HYPOTHESES

OBJECTIVES: After studying this chapter and working the problems, you should be able to:

Explain that hypothesis testing is something we all do, statisticians or not.

Set up a testing situation that uses the null or alternative hypothesis concepts.

Demonstrate the consequences of making type I and type II errors, and illustrate how α and β affect them.

Demonstrate how to make calculations from sample data.

Make the appropriate statistical decision.

A HYPOTHESIS IS nothing more than a guess or an assumption. Even though you are not aware of it, you are constantly making hypotheses, testing them, and then accepting or rejecting them. For instance, when you awoke this morning, you made a hypothesis: "It is worthwhile to get up." You then took in some information—what kind of day is it? how do I feel? how bad do I look?—and finally decided to accept the hypothesis. From then on, you are in a constant process of making and testing hypotheses: I need a cup of coffee; I have time for a second cigarette before I leave for work; I can make it across this intersection before that car gets here; my boss won't care if this report is late. And so on through the day. Notice one important factor in each of these situations—the decision about accepting or rejecting a hypothesis is made before all the final information is in your hands.

In other words, you don't know for sure whether it's worthwhile to get out of bed until you've done it. There's always the possibility of making the wrong decision.

Let's put this notion in the context of formal statistical hypothesis testing. A statistical hypothesis is an assumption about a parameter. Hypothesis testing helps us make decisions about uncertain situations. If our hypothesis is "it will rain today" and we look out the window and see it raining, it is not very difficult to accept the hypothesis. But, if it is merely lightly overcast outside, we have a harder time making a decision. Statistical hypothesis testing helps us in the "gray" areas and, conversely, would be totally superficial in "black or white" situations where the answer is obvious.

Suppose you are buying two-inch aluminum pipe to run water to a drilling rig. The supplier is selling to you (and charging to you) 30-foot sections of this pipe. A sample of 100 of the 30-foot sections shows an average length of 20 feet. Do you accept the hypothesis that the average is at least 30 feet? Of course you don't. What if the average of the sample of 100 is 25 feet? 28 feet? 29.5 feet? 29.9 feet? You can begin to see that there is some area in which you are uncertain about the answer. If the average length is 29.9 feet, you are obviously not going to take the supplier to court.

You are beginning to recognize that there may be some error in your sample. If the sample of 100 pipes has an average length of 28 feet, do you sue the supplier or do you believe that the population mean really is 30 feet and you just happened to draw a sample of predominantly short pipes? This type of problem can be handled through a test of hypothesis. Concerning the pipe lengths, we are questioning the average of the entire population of pipe lengths. In particular, we would hypothesize here that $\mu \geq 30$ feet. Now the only way we can be totally certain that the average length is or is not at least 30 feet is to measure all the pipe in the population. If the population is very large (which this one is), this total census would be infeasible.

Generally we are left with making an inference about the population mean based on our knowledge of sample data. When we make inferences, we must recognize that we can make an error. We want to examine or test our hypothesis about the average pipe length and "control" or at least understand the odds that we may make an erroneous inference.

A test of hypothesis, like other statistical techniques, is not magically going to yield a guaranteed-correct answer. In an uncertain world there is no chance of coming up with such a thing. But if we recognize that the world is uncertain, the next best thing is to control or know the probability of making an error. A statistical test of hypotheses does this for us. To put this problem in a formal test of hypothesis, we follow a seven-step procedure.

Step 1: Formulate a *null* hypothesis and an *alternative* hypothesis. In our situation these would be

Null hypothesis: H_0: $\mu \geqslant 30$ feet

Alternative hypothesis: H_1: $\mu < 30$ feet

There are a number of ways of determining the null hypothesis. From experience I believe the best way is to ask yourself, "What do I believe without any sample data?" The null hypothesis is the answer to this question. In our case we certainly can't claim that the supplier is cheating us without any sample information to support us. Therefore we must believe that the true mean length is at least 30 feet as claimed. The alternative hypothesis is what we will believe if we reject the null hypothesis.

Step 2: Determine the level of significance. There are four different possibilities of outcomes when we test a hypothesis, as shown in Table 10-1.

TABLE 10-1. Possible Outcomes of Hypothesis Testing

	Null Hypothesis Is:	
	Accepted	Rejected
Null hypothesis is true	No error	Type I error
Null hypothesis is false	Type II error	No error

If we accept the null hypothesis when it is true or if we reject the null hypothesis when it is false, we have obviously made the correct decision. But if we reject the null hypothesis when it is true, we have made a *type I error*. The probability of making a type I error is α: Pr(type I error) $= \alpha$; α is the level of significance. The level of significance, one of the easily controlled items, is the percentage of time, over a long period, that the analyst will make a type I error using this criterion. Normally α is set at .10, .05, .02, or .01. For our example, let $\alpha = .05$.

Looking at Table 10-1, we can logically conclude that unless we take a complete census, the only way we can eliminate the possibility of making a type I error is by *always* accepting the null hypothesis. This, of course, will vastly increase our chances of making a type II error. Similarly, if we *always* reject the null hypothesis there is no way we can ever make a type II error—but this policy will wreak havoc on our chances of making a type I error. We denote the probability of a type II error by β: Pr(type II error) $= \beta$. If the sample size is kept constant, β will generally decrease as α increases. Conversely, β will in-increase as α decreases. Both α and β can be decreased simultaneously by increasing the sample size. At the extreme end, both α and β would be zero if we sampled the entire population (a complete census).

Step 3:   Determine the best testing statistic to be used. In our example, the central limit theorem tells us that the means of samples of size 100 tend to be normally distributed. Therefore our testing statistic (SO&B 8-1) is

$$z = \frac{\bar{x} - \mu}{\sigma_{\bar{x}}}$$

Step 4:   Find the region of rejection (the RR). The major purpose of a formal test of hypothesis is to identify a decision point or points. If the sample result is on one side of the decision point, we accept the hypothesis; if the sample result is on the other side, we reject the hypothesis. We want to make this decision point so that we can identify the probability

FIGURE 10-1.   Region of Rejection

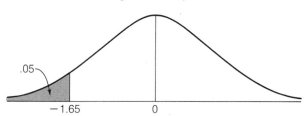

of making a type I error (α). The region of rejection tells us explicitly where we will reject the null hypothesis. This region is uniquely determined by steps 1, 2, and 3. If the random variable is normally distributed (step 3), then 5 percent (step 2) of the time z will be less than -1.65. The region of rejection is shown in Figure 10-1.

In other words, when we reject the null hypothesis for values of z less than -1.65, the null hypothesis would be true 5 percent of the time even though it was rejected. This represents, of course, a 5 percent chance of making a type I error. We write the region of rejection:

$$\text{RR: } z < -1.65$$

Note that we have done the first four steps without knowing any sample data. And this is as it should be, since knowledge of sample results can introduce a subconscious bias in setting up our test.

Step 5: Make the sample calculations. Suppose in our example that the sample of 100 yields a mean $\bar{x}$ of 29 and a standard deviation S of 5. The sample is large enough for us to say that $S \simeq \sigma$. We can now calculate our testing statistic:

$$z = \frac{\bar{x} - \mu}{\sigma_{\bar{x}}} \quad (\mu \text{ is our hypothesized } \mu)$$

$$H_0 \geqslant 30$$

$$= \frac{29 - 30}{5/\sqrt{100}} \quad \sigma_{\bar{x}} = \frac{S_D}{\sqrt{100}} \quad \frac{S_D}{\sqrt{N}}$$

$$= \frac{-1}{.5}$$

$$= -2$$

The z value of -2 lies in the region of rejection. Therefore we reject the null hypothesis. Rejection of the null hypothesis automatically implies acceptance of the alternative.

Step 6: Make the statistical decision. In our example this is "reject the null hypothesis."

Step 7: Take the management action required. This would be to find a new supplier.

Notice in our example that if we set $\alpha = .01$, then the region of rejection is $z < -2.33$ (from Table A-3 in the appendix). The null hypothesis would not be rejected because the z value from step 5 is not in that region of rejection. If we want to lower our risk (from .05 to .01) of rejecting a true hypothesis, we will reject over a narrower range. If it is harder to reject a true hypothesis, though, it will be easier to accept a false one.

ONE-TAILED VERSUS TWO-TAILED TESTS

The example concerning the 30-foot lengths of pipe was a *one-tailed test*. It is called that because the entire region of rejection is in one tail. The one tail contains all this region because of the null hypothesis. We were interested only in the lower side of the distribution and really didn't care how far above 30 feet the actual mean μ may have been. It could have been 100 feet for all we cared.

Suppose, though, that we are now the manufacturer of this aluminum pipe. We are still concerned if the mean is significantly below 30 feet because our customers will complain. But we are equally concerned if the mean is greater than 30 feet because we would then be losing money (we're only charging for 30 feet). We would therefore want to take some action if our pipe lengths were too long. Hence if we are the manufacturer, we will test as follows:

1. $H_0: \mu = 30$ feet
       $H_1: \mu \neq 30$ feet

2. $\alpha = .05$

3. $z = (\bar{x} - \mu)/\sigma_{\bar{x}}$

4.   RR: $z < -1.96$ or $z > 1.96$

The region of rejection is now in two tails as shown in Figure 10-2. The *sum* of the two tails is our level of significance, $\alpha = .05$. Put another way, in a *two-tailed test* we split our level of significance into two equal segments. We still have a 5 percent chance of rejecting the hypothesis when it is true. To continue with our test:

5. $z = (\bar{x} - \mu)/\sigma_{\bar{x}} = -2$ (as before)

6.   Reject the null hypothesis.

7.   Look for an assignable cause of the production system being out of kilter.

FIGURE 10-2. Two-tailed Region of Rejection

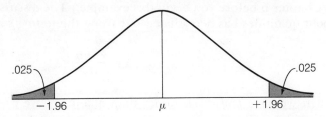

.025   .025

-1.96 μ $+1.96$

Example 1: Suppose we are receiving a large shipment of vials of serum. These vials are supposed to contain 100 mg of serum each, and we would like to test the hypothesis that the average vial really contains 100 mg. Because we are just as concerned about the vial containing too much serum as too little, we will make it a two-tailed test. Let's set α at .01. A sample of 64 vials shows a mean content ($\bar{x}$) of 99.25 mg. We also know that the variance of the contents is 4 mg². (Therefore the standard deviation would be $\sqrt{4}$, or 2.) The solution would be:

1. H_0: $\mu = 100$ mg
   H_1: $\mu \neq 100$ mg

2. $\alpha = .01$

3. $z = \dfrac{\bar{x} - \mu}{\sigma_{\bar{x}}}$

4. RR: $z > 2.57$ or $z < -2.57$

5. $z = \dfrac{99.25 - 100}{2/\sqrt{64}} = -\dfrac{.75}{.25} = -3$

6. Reject the null hypothesis.

7. Return the shipment for not meeting your specifications—knowing that 1 of every 100 times you conduct the acceptance experiment in this manner you will be returning good vials.

SAMPLE PROPORTION ($\bar{p}$) VERSUS POPULATION PROPORTION (P)

Suppose we are concerned about proportions rather than quantities. In testing a hypothesis about a proportion, we call upon the normal approximation to the binomial. (You may want to review that section in Chapter 6 before you begin this example.) The discussion thus far about quantities has necessitated our using the testing statistic

$$z = \frac{\bar{x} - \mu}{\sigma_{\bar{x}}}$$

When we have *proportions* rather than quantities, we use the testing statistic from SO&B 6-5:

$$z = \frac{\bar{p} - P}{\sqrt{PQ/n}}$$

Suppose we are concerned about the outcome of an upcoming election. Our population N is the total number of people who will cast a vote. We are interested in estimating the outcome before the election, though. It would, of course, be totally infeasible to find each person who is going to vote and determine exactly how he will vote. Therefore we will satisfy our curiosity with a sample survey.

We are particularly interested in Mabel Metcher, who is actively campaigning for president of her association, The Mismatchers. Thus far Mabel's campaign has been going smoothly. As the election draws near, her campaign committee decides to take a sample of 100 voters of this large organization and estimate the proportion that will vote for Mabel. The committee believes that Mabel will win the election. In other words, they are hypothesizing that at least 50 percent of the voters favor her.

Let's review the concept of hypothesis testing. What if no one, in the sample of 100, was going to vote for Mabel? Would you still believe she'll win the election? Of course not. Would you still believe the hypothesis if 49 percent of the sample favored her? Probably. At least you would *not* be willing to reject the hypothesis and urge her to quit the campaign. The point is that there is some proportion above which you will not reject the hypothesis and below which you would say the difference is too big to be merely sampling error and reject the hypothesis. The test of hypothesis will tell us what that point is.

In the sample survey of 100 voters, 41 are found to favor Mabel. The formal test of hypothesis would be:

1. H_0: $P \geq .5$
   H_1: $P < .5$

2. $\alpha = .025$

3. $z = \dfrac{\bar{p} - P}{\sqrt{PQ/n}}$

4. RR: $z < -1.96$

5. $z = \dfrac{\bar{p} - P}{\sqrt{PQ/n}} = \dfrac{.41 - .50}{\sqrt{(.5)(.5)/100}} = -1.80$

6. Do not reject H_0.

7. Continue the campaign.

PR(TYPE II ERROR)

We control the level of significance α in a test of hypothesis very specifically (by step 2). But β, the probability of type II error, is also of importance. As α increases, β decreases; and as α decreases, β increases when the sample size is kept constant. We can decrease α and β simultaneously by increasing the sample size.

To illustrate the concept of a β error, let's look at two examples concerning our friend Mabel Metcher. Mabel often meets her friend Finley for drinks and at Finley's suggestion, they flip a coin to see who pays for the drinks. Mabel notices that Finley always flips the coin and he always calls heads. She also notices that she usually ends up paying for the drinks. Mabel begins to suspect Finley of cheating. She knows the coin is not two-headed, but she also knows there is a counterfeit coin that will come up heads three-fourths of the time rather than one-half of the time. Does Finley have such a counterfeit coin? Mabel decides to let Finley flip the coin just one more time. If it comes up heads again, she will assume he is cheating and will never see him again. This will of course end a beautiful friendship. If, on the other hand, the coin comes up tails, she will assume he has a fair coin.

Mabel, in this experiment, is hypothesizing that the coin is fair. Her null and alternative hypotheses concerning the proportion of heads would be:

$$H_0: P = \tfrac{1}{2} \quad \text{(the coin is fair)}$$

$$H_1: P = \tfrac{3}{4} \quad \text{(the coin is counterfeit)}$$

The level of significance (the probability of rejecting the null hypothesis when it is true) is the probability of getting a head on a fair coin;

hence $\alpha = \frac{1}{2}$. Finley should be very concerned that this level is so high.

The β error (the probability of accepting the null hypothesis when it is false) is the probability of getting a tail on a counterfeit coin. Remember that Mabel will accept the null hypothesis (fair coin) if a tail comes up. Therefore β in our example would be $\frac{1}{4}$, since the probability of a tail on the counterfeit coin is $\frac{1}{4}$.

The *power* of a test is defined as the probability of rejecting a hypothesis when it is false, or $1 - \beta$. The power of Mabel's test is $1 - \frac{1}{4} = \frac{3}{4}$. When presented with a choice between tests, you would choose the one with the highest power, all other things being equal.

THE ELECTION TYPE II ERROR

Looking back at Mabel's election campaign we can calculate the probability of making a type II error in that situation. To calculate a β we must have a specific value in the alternative hypothesis. Instead of H_1: $P < .50$, let's make H_1: $P = .40$. What is the value of β for this situation?

Figure 10-3 shows the situation under the condition that the null hypothesis is true. The shaded area is α. The curve beneath this shows the situation under the condition that the alternative hy-

FIGURE 10-3.   The Value of β

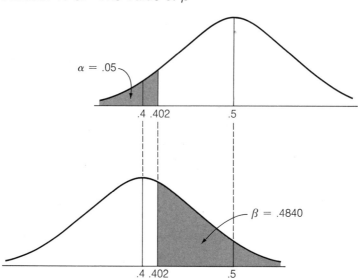

pothesis is true, or $P = .40$. Notice in the lower curve that the shaded area represents the *null* hypothesis being accepted, even though it is false. Note also that the value .402 is merely our region of rejection translated back to absolute units: $.5 - 1.96(.05) = .5 - .0980 = .402$.

We have already set the probability of rejecting the null hypothesis when it is true at .025. To calculate β we merely go back to our normal distribution:

$$\Pr(\bar{p} \geq .402 | n = 100, P = .4)$$

$$z = \frac{.402 - .4}{\sqrt{(.4)(.6)/100}}$$

$$= \frac{.002}{.049}$$

$$= .041$$

Hence from our normal tables we find that

$$\beta = \Pr(\bar{p} \geq .402 | n = 100, P = .4)$$

$$= .4840$$

In other words, if we want to limit our chance of rejecting H_0 to .025, then we are running a 48 percent chance of accepting H_0 when the true proportion is only $P = .4$. If we performed this calculation for each possible value of P for an alternative hypothesis, we would end up with what is called an *operating characteristic* (OC) curve.

The astute reader (in other words, the guy who is always complaining, "That isn't what you said yesterday") will have noted that when we discussed the normal approximation to the binomial, I made a big point of a correction factor for transferring from a discrete real world to a continuous normal curve. This correction factor has been ignored here because it will be quite small with a sample of 100 and because the practical user, for whom this book is written, will not need it unless he finds his final results to be very close.

AUTOPSY

We have looked at the basic concept of a test of hypothesis and have examined the two decisions of "accept" or "reject" the hypothesis. Actually the action of rejecting a hypothesis is more conclusive than not rejecting because we know the probability of making an error when we reject the hypothesis. This would be the probability of making a type I error or α, which we determined in step 2. We do not know, without additional calculations, the probability of a type II error, which would be associated with accepting the hypothesis. To be technically correct, we should either reject the null hypothesis or reserve judgment. However, in practical, nonresearch situations acceptance is implied rather than the technically correct decision — reserve judgment.

Statistical techniques were designed and used originally in pure research, where the consequence of making a type I error was harsh. Hence most statistical studies still use an α of .01 or .05. As we discussed, this usually results in a high β level. In business, though, the consequence of making a type I error may not be much worse than making a type II error. Consequently, I have successfully used α levels of .10 or even .20 in real business situations. These higher α levels substantially reduce the β level, the probability of accepting a null hypothesis when it is false.

SUMMARY

The Seven Steps of Hypothesis Testing

1. Determine the null and alternative hypotheses.

2. Determine the level of significance.

3. Determine the testing statistic.

4. Find the region of rejection.

5. Make calculations from sample data.

6. Make the statistical decision.

7. Take management action if appropriate.

The hypothesis test is *one-tailed* if we are only concerned about the parameter being larger than a certain value or we are only concerned about a parameter being smaller than a certain value.

The hypothesis test is *two-tailed* if we are concerned about the value of a parameter being either larger than or smaller than a certain value.

Important Terms

Type I error: rejecting a null hypothesis when it is true

Type II error: accepting a null hypothesis when it is false

Pr(type I error) $= \alpha$

Pr(type II error) $= \beta$

Power of a test $= 1 - \beta$

PROBLEMS

1.   Think of three examples of informal hypothesis testing that you have done today. What were the null and alternative hypotheses? Did you make a wrong decision in any of them? Was it a type I or a type II error?

2.   A manufacturer of dry cereal is producing 20-ounce packages of his product. The automated packaging device needs frequent checking to see whether it is actually putting 20 ounces in each package. The weights of the packages are known to be normally distributed with a variance of .25 ounce². A sample of 49 packages shows an average weight $\bar{x}$ of 19.8 ounces. Test the appropriate hypothesis at a 5 percent level of significance. (*Hint:* The manufacturer will be equally upset if the machine is overfilling or underfilling; therefore the test must be two-tailed.)

3.   A trucking company has kept a close and accurate record of tire wear on their trucks. This voluminous history shows that the number of miles driven before a tire needs replacement is roughly normally distributed and has a standard deviation of 400 miles. The tires they have been using last an average of 20,000 miles. A new brand is being offered to them with the claim that it will last longer. They buy 100 of these tires and find that the average mileage before replacement is 20,100 miles. At an α level of 10 percent, would you conclude that the new brand is better?

4.   A manufacturer of hydraulic hose has one product with a standard bursting strength of 2000 pounds per square inch (psi). The manufacturing process has been going for a long time and the chief of production knows that the bursting strength of this hydraulic

hose is normally distributed with a standard deviation of 20 psi. A routine random sample of 50 pieces is taken to the laboratory for testing. The average bursting strength of these 50 pieces is found to be 1995.4 psi. Test the hypothesis that the process is making hoses with at least 2000-psi bursting strengths. Use a level of significance of .05.

5.   Suppose you are the proprietor of a gift shop that carries low-cost, impulse-purchase merchandise. The average sales ticket is for $5.63 worth of merchandise. A long history of these sales shows that $\sigma = 1.40$. A salesman convinces you that by installing air conditioning in your shop, the average sales ticket will go up. A random sample of 49 sales tickets after the air conditioning is installed shows an average of $5.95. Should you conclude you were correct in installing the air conditioning? Use an α of .10. Why could you use the normal curve?

6.   An auditor is working for a person who is considering purchasing a defunct business. The defunct business has stated that no more than 30 percent of its accounts receivable are more than 30 days past due. The auditor takes a random sample of 64 accounts and finds that 24 of them are more than 30 days past due. Test the appropriate hypothesis at an α level of .025.

7.   Little Red Riding Hood has complained to the woodsman about the Big Bad Wolf's activity. The fair and honest woodsman has consequently reached an agreement with the wolf that he must not always be attacking L.R.R.H. The wolf agrees to a contract whereby he will attack L.R.R.H. 50 percent of the time and she will attack him 50 percent of the time. On the next 40 trips through the woods, the wolf attacks L.R.R.H. 24 times. The woodsman now wants to test the

hypothesis that the wolf is living up to his contract. Use an α level of .01.

8. A new pen has been developed by a firm that will offer a full money-back guarantee if the product is defective. The firm can produce these pens very cheaply if they allow much of their production to be defective. The marketing department claims that the product will be quite profitable because fewer than 40 percent of the customers who get a defective pen will bother to come back for their refund. The accounting department, though, feels that the profit will be only marginal because they estimate that 45 percent of the defective pens will result in a demand for refund. The marketing research department conducts a survey of 100 people and finds 44 who would demand a refund.

> *a.* Test the hypothesis that fewer than 40 percent will demand refunds against the alternative hypothesis that $P = .45$.
>
> *b.* Find the value of β for this problem at an α level of $10, 5$, and 1 percent.

9. A politician has publicly directed charges against the state welfare department claiming that "one out of every four recipients of welfare involves fraud." An in-depth investigation of a sample of 100 recipients is conducted. The results show that five cases in this sample involved some kind of fraud. Test the hypothesis that the proportion of fraud is equal to or greater than 25 percent at a level of confidence of .05.

10. A standardized aptitude test for a certain trade shows that the nationwide scores reflect an average of 185 and a standard deviation of 16. A sample of 64 graduates of one training school shows an average grade of 182. At a confidence level of .05, can you conclude that the graduates of this training school do significantly worse than the national average?

SELECTED REFERENCES

Campbell, Stephen K. *Flaws and Fallacies in Statistical Thinking.* Englewood Cliffs, N.J.: Prentice-Hall, 1974. Especially Chapters 10, 12.

Dixon, W. J. and F. J. Massey. *Introduction to Statistical Analysis.* Third edition. New York: McGraw-Hill, 1969.

Guenther, William C. *Concepts of Statistical Inference.* Second edition. New York: McGraw-Hill, 1973. Especially Chapter 4.

Tanur, Judith M., Frederick Mosteller, William H. Kruskal, Richard F. Link, Richard S. Pieters, and Gerald R. Rising. *Statistics: A Guide to the Unknown.* San Francisco: Holden-Day, 1972.

Yamane, Taro. *Statistics: An Introductory Analysis.* Third edition. New York: Harper & Row, 1973. Especially Chapter 9.

CHAPTER 11
MORE TESTS
OF HYPOTHESES

OBJECTIVES: After studying this chapter and working the problems, you should be able to:

Explain when to use t *instead of z.*

Demonstrate how to use the t *test for determining significant differences between:*

   A sample mean and a hypothesized population mean

   Two means of independent samples

   Two means of dependent samples

Decide when to use chi square, *and apply it to a test of a hypothesis about a sample variance.*

Demonstrate the appropriate use of the F *test.*

Demonstrate how analysis of variance *can be used to compare the means of several different samples.*

IN CHAPTER 10 we covered the basic ideas of testing a hypothesis. There are many different situations in which a hypothesis to be tested can arise. We will now expand our knowledge to cover some other areas. Keep in mind that these formal tests of hypothesis are ways to find a point of delineation within a gray area—an area in which we are uncertain about the meaning of the outcome.

THE *t* DISTRIBUTION

The tests of hypotheses presented in Chapter 10 were in situations where we could apply the normal distribution. In other words, either we knew σ or else the sample size was large enough for us to rely on the central limit theorem. We often run into the situation, though, in which we want to test a hypothesis concerning a mean but our sample size is too small to be approximated by the normal distribution. In this situation, referred to by some statisticians as small-sample testing, we can use the *t distribution*. The *t* distribution is a symmetric, bell-shaped distribution that approaches the normal distribution as *n*, the sample size, increases.

We know (at least I know it) that if a sample of size *n* is drawn from a normal population, then

SO&B 11-1: $$t = \frac{\bar{x} - \mu}{S_{\bar{x}}}$$

follows the *t* distribution with $n - 1$ degrees of freedom. (I'll explain "degrees of freedom" in a moment.) This distribution is similar in form to the standardized normal,

$$z = \frac{\bar{x} - \mu}{\sigma_{\bar{x}}}$$

except that here we are working with the standard deviation of the sample, *S*, rather than the known population standard deviation σ. As a practical rule of thumb, the *t* distribution is generally used when the sample size *n* is less than 30. Now let's go back to degrees of freedom.

Degrees of freedom, a fundamental concept, can be thought of in the same light as adding five numbers together when you know their sum:

$$
\begin{array}{r}
5 \\
2 \\
1 \\
3 \\
\underline{x} \\
15
\end{array}
$$

Obviously the missing number has to be 4. In other words, we have freedom of choice for the first four digits, but the fifth digit is locked in once we know the other four. We might say, then, that this example has 4 degrees of freedom. In concept, then, a column of *n* numbers adding to a known sum has $n - 1$ degrees of freedom (df).

A table of the t distribution is presented in Table A-4 in the appendix. To illustrate the use of this table, suppose we look for the t value associated with a sample size n of 25 and a one-tailed α level of .05. In Table A-4 we would read across the first row (one tail) to .05 and then follow that column down to the proper degrees of freedom, which is $n - 1 = 25 - 1 = 24$. We should arrive at a t value of 1.711. If we were interested in a two-tailed α level of .05, which would put .025 in each of the two tails, we would read across the second row (two tails) to .05. We would follow this column down to 24 degrees of freedom to find a t value of 2.064. Notice that at the bottom of the table, with infinite degrees of freedom, the t values are the same as the corresponding z values for the normal distribution. This fact illustrates that the t distribution approaches the normal distribution as n increases.

We can use the t distribution to test hypotheses concerning means when we do not know σ, the sample size is less than 30, and the variable of concern is approximately normally distributed. We still follow the seven-step procedure for testing a hypothesis. In step 3, though, we use the testing statistic

$$t = \frac{\bar{x} - \mu}{S_{\bar{x}}} \quad \text{with } n - 1 \text{ df}$$

instead of

$$z = \frac{\bar{x} - \mu}{\sigma_{\bar{x}}}$$

From this point we will use "df" to denote degrees of freedom.

Example 1: Suppose we are the manufacturer of wire cable. One of our products is sold and advertised as having a breaking strength of 200 pounds per square inch (psi). A random sample of nine sections of this product shows a mean breaking strength of 212 psi and a standard deviation of 12 psi. We are concerned about the process being at 200 psi. Past experience leads us to believe that the breaking strength of cable is approximately normally distributed. This will be a two-tailed test because we will have customer complaints if our product is lower than standard and we are probably spending too much money on manufacturing if our product is above standard. Note that σ is unknown and $n < 30$. We would proceed as follows:

1. H_0: $\mu = 200$
   H_1: $\mu \neq 200$

2. $\alpha = .05$ (chosen subjectively)

3. $t = (\bar{x} - \mu)/S_{\bar{x}}$ with $n - 1$ df

4. RR: $t > 2.306$ or $t < -2.306$ (from our tables of the t distribution with $9 - 1 = 8$ df)

5. $t = \dfrac{212 - 200}{12/\sqrt{9}} = \dfrac{12}{4} = 3$

6. Reject H_0.

7. Try to find our why the process is out of kilter.

Example 2: Supposeth there is a Danish prince who is wondering whether he should go on suffering the slings

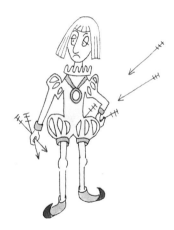

and arrows of outrageous fortune or simply do himself in. In addition to being melancholy, this prince is also versed in statistical methods. He decideth that he could really suffer an average of 20 slings and arrows per day and still continue his life. But if he must suffereth significantly more than that, he will decide tis nobler to end it right here and noweth.

The prince carefully chooses a random sample of past days and finds that he suffered the following slings and arrows:

Day	Slings and Arrows Suffered
1	25
2	16
3	28
4	21
5	19
6	22

He further feels that it is safe to assume the slings and arrows per day follow approximately a normal distribution. He then maketh the following test of hypothesis:

1. H_0: $\mu \leqslant 20$
    H_1: $\mu > 20$

2. $\alpha = .05$

3. $t = \dfrac{\bar{x} - \mu}{S_{\bar{x}}}$

4. RR: $t > 2.015$

5. $\bar{x} = \dfrac{\Sigma x}{n} = \dfrac{131}{6} = 21.83$

$$S = \sqrt{\dfrac{n\Sigma x^2 - (\Sigma x)^2}{n(n-1)}}$$

$$= \sqrt{\dfrac{6(2951) - (131)^2}{6(5)}}$$

$$= \sqrt{\dfrac{545}{30}} = \sqrt{18.17} = 4.3$$

$$S_{\bar{x}} = \dfrac{4.3}{\sqrt{6}} = 1.52$$

$$t = \dfrac{21.83 - 20}{4.3/\sqrt{6}} = 1.20$$

6. Do not rejecteth the null hypothesis ($1.20 < 2.015$).

7. Do not yell uncle—if you'll excuse the expression—here.

You will find it beneficial to verifieth *all* the figures in step 5.

Example 3:   A consumer feels that the weights printed on packaged meat at the supermarket are exaggerated. He takes ten packages and weighs them under laboratory conditions. These ten observations show the following deviations from the true weight:

$$-.1, 0, +.4, +.2, 0, +.3, -.2, +.1, +.2, -.1$$

The consumer feels that the deviations from true weight are approximately normally distributed. The test of hypothesis should look like this:

1. H_0: $\mu \leqslant 0$
    H_1: $\mu > 0$

Note that the consumer must assume the store's weights are equal to or less than true and can only state otherwise if he can find sample data to reject that null hypothesis.

2. $\alpha = .05$

3. $t = (\bar{x} - \mu)/S_{\bar{x}}$ with 9 df

4. RR: $t > 1.833$

5. From the sample data shown,

$$\bar{x} = \frac{.80}{10} = .08$$

$$S = \sqrt{\frac{10(.40) - (.80)^2}{10(9)}} = \sqrt{.037} = .192$$

$$t = \frac{.08 - 0}{.192/\sqrt{10}} = \frac{.08}{.061} = 1.31$$

6. Do not reject H_0.

7. Do not try to press charges against the supermarket.

THE TEST FOR DIFFERENCE BETWEEN TWO MEANS— INDEPENDENT SAMPLES

Thus far we have concerned ourselves with testing a sample mean against some hypothesized population mean. Often, however, we have two sample means and want to know whether there is a significant difference between them. If the two samples are drawn from normal distributions and the variances of the two populations can be assumed to be equal, we can use the testing statistic

SO&B 11-2:
$$t = \frac{(\bar{x}_1 - \bar{x}_2) - (\mu_1 - \mu_2)}{\sqrt{\dfrac{(n_1 - 1)S_1^2 + (n_2 - 1)S_2^2}{n_1 + n_2 - 2}}\sqrt{\dfrac{n_1 + n_2}{n_1 \cdot n_2}}}$$

with $n_1 + n_2 - 2$ df. The term

$$\sqrt{\frac{(n_1 - 1)S_1^2 + (n_2 - 1)S_2^2}{n_1 + n_2 - 2}}$$

is an estimate of the "pooled" or combined standard deviation of the two samples. This pooled standard deviation is usually denoted by $S_{x_1-x_2}$.

To illustrate the use of testing for differences between two independent means, imagine that you are the sales manager for a chain of stores that sell only men's ties. While planning a good advertising campaign, you begin to wonder who spends more: male customers shopping for themselves or female customers shopping for a man. You decide to observe a sample of customers and record the amount of their purchase classified as to whether the purchaser is male or female. Here are the results of that sample:

Male Customers	Female Customers
$\bar{x}_m = \$7.30$	$\bar{x}_f = \$8.20$
$S_m^2 = 1.44$	$S_f^2 = 1.69$
$n_m = 15$	$n_f = 12$

As a quick review of the reason for testing a hypothesis, we cannot automatically assume from the sample results that female customers spend more per purchase than males. We need to know whether the result from the sample is *significant*. We ask ourselves, "Self, what would I believe to be true if I had no data?" The answer would have to be, "There is no difference in the average value of purchases of men versus women." Therefore our test of hypothesis becomes:

1. $H_0: \mu_m = \mu_f$ or $\mu_m - \mu_f = 0$
    $H_1: \mu_m \neq \mu_f$ or $\mu_m - \mu_f \neq 0$

Remember that sample data should not influence your choice of hypotheses.

2. $\alpha = .10$

3. $t = \dfrac{(\bar{x}_m - \bar{x}_f) - (\mu_m - \mu_f)}{\sqrt{\dfrac{(n_m - 1)S_m^2 + (n_f - 1)S_f^2}{n_m + n_f - 2}}\ \sqrt{\dfrac{n_m + n_f}{n_m \cdot n_f}}}$

with $n_m + n_f - 2$ degrees of freedom. Remember: We are assuming that the sample came from normal distributions and the variances of the two populations are equal. (We will discuss how to determine equality of variances in a later section.)

4. RR: $t > 1.708$ or $t < -1.708$

5. $t = \dfrac{(7.30 - 8.20) - 0}{\sqrt{14(1.44) + 11(1.69)/25}\ \sqrt{(15 + 12)/(15 \cdot 12)}}$

    $= \dfrac{-.90}{\sqrt{1.55}\ \sqrt{.15}}$

    $= \dfrac{-.90}{\sqrt{.233}} = \dfrac{-.90}{.48} = -1.87$

6. Reject the null hypothesis.

7. Direct at least some of your advertising to female shoppers.

THE TEST FOR DIFFERENCE BETWEEN TWO MEANS — DEPENDENT SAMPLES

When we are concerned about the difference between two sample means, we can be working with either independent samples or dependent samples. With independent samples there is no tie between the observations. For instance, if we are trying to find out whether there is a significant difference between two different tests, we could give test A to one group of people and test B to a second group. The samples are independent because the two groups have nothing to do with each other. We would analyze the difference between the mean scores of the two groups by using SO&B 11-2:

$$t = \frac{(\bar{x}_1 - \bar{x}_2) - (\mu_1 - \mu_2)}{S_{x_1-x_2} \sqrt{(n_1 + n_2)/n_1 \cdot n_2}}$$

from the preceding section.

However, a problem could arise here. After we've analyzed the two sample means, we don't know whether we have measured the difference in the tests or the difference in the two groups of people. This doubt can be overcome by making the experiment with *dependent* samples—that is, giving both tests to the same group of people. If we administer the tests so that each person takes both tests, then the two samples are tied together: they are dependent in that we now have two results from the *same* person. We no longer have to worry about the difference in ability of the test taker because it's the same person.

Instead of working with two different samples, we create the difference in the two tests for each test taker. These differences will then be treated like any other group of data for hypothesis testing. To illustrate let's put some numbers in our test-taking example. There are two tests available that supposedly measure a person's mathematical ability. Test A is fairly well established as being accurate. Test B is new, but is much easier to administer. We would like to see whether the two tests yield the same result. We give the two tests to a group of ten people with the following results:

Test Taker	Results Test A	Test B	Difference (A − B)	(A − B)²
1	90	92	−2	4
2	70	69	1	1
3	84	88	−4	16
4	65	70	−5	25
5	60	55	5	25
6	50	50	0	0
7	65	66	−1	1
8	74	72	2	4
9	95	96	−1	1
10	85	90	−5	25
			−10	102

The column of differences (A − B) becomes our only point of concern. Let each individual difference in the column be denoted by d_i. We find the average difference $\bar{d}$ in the usual manner:

$$\bar{d} = \frac{\Sigma d_i}{n}$$

Similarly,

$$S_d = \sqrt{\frac{n\Sigma d_i^2 - (\Sigma d_i)^2}{n(n-1)}}$$

In our example concerning the two tests,

$$\bar{d} = \frac{\Sigma d_i}{n} = \frac{-2 + 1 - 4 - 5 + 5 + 0 - 1 + 2 - 1 - 5}{10}$$

$$= \frac{-10}{10}$$

$$= -1$$

Notice that you must keep track of the sign of each individual difference.

$$S_d = \sqrt{\frac{10(102) - (-10)^2}{10(9)}}$$

$$= 3.19$$

Now we are concerned with that familiar-sounding question, "Is the average difference in the two tests significantly different from zero?" Our formal test of hypothesis, using the testing statistic SO&B 11-3,

SO&B 11-3: $$t = \frac{\bar{d} - \mu_d}{S_{\bar{d}}}$$

would be this:

1. $H_0: \mu_d = 0$ (μ_d is the true mean difference in the populations)

    $H_1: \mu_d \neq 0$

2. $\alpha = .10$

3. $t = (\bar{d} - \mu_d)/S_{\bar{d}}$ with $n - 1$ df

4. RR: $t > 1.833$ or $t < -1.833$

5. $t = \dfrac{-1.0 - 0}{3.19/\sqrt{10}} = \dfrac{-1.0}{1.009} = -.99$

6. Do not reject H_0.

7. Use test B if it is easier to administer.

I hope you have now grasped the difference between an independent sample test and a dependent sample test. At the risk of boring you, I will once more iterate: With independent samples, the analyst has to be certain that the variable he is measuring is the only variable involved between the two samples. This is why I recommend that you first try to get the experiment in a dependent form. The dependent form reduces the chance of measuring some extraneous variable.

Let's look at one more example of a dependent sample before we conclude the subject. The Booze Before Breakfast Group meets every morning for a couple of quickies before work, and because of this habit they tend to have trouble holding their jobs. Now the bartender is concerned about this tendency because whenever they are not receiving a paycheck, they try to get drinks on the house. He suggests that they drink vodka instead of bourbon to conceal their habits from the boss. Eight drinkers volunteer to try his plan. The bartender wants to get controlled results of the experiment; fortunately, his other job is as professor of statistics at State University around the corner. He records for each volunteer the number of days he held his last job. He then records the number of days each holds his new job after switching to vodka.

Here we are only concerned about an increase and, of course, we cannot assume that there is any increase before we have sample data, so our hypotheses and the formal test would become:

1. $H_0: \mu_d \leq 0$
   $H_1: \mu_d > 0$

2. $\alpha = .10$

3. $t = (\bar{d} - \mu_d)/S_{\bar{d}}$ with $8 - 1 = 7$ df

4. RR: $t > 1.415$

5. Our experiment shows the following results:

Volunteer	Days Held Job Vodka	Bourbon	d_i	d_i^2
1	25	20	+5	25
2	18	18	0	0
3	24	22	+2	4
4	17	19	−2	4
5	25	21	+4	16
6	16	17	−1	1
7	21	14	+7	49
8	14	13	+1	1
			16	100

$$\bar{d} = \frac{\Sigma d_i}{n} = \frac{16}{8} = 2$$

$$S_d = \sqrt{\frac{n\Sigma d_i^2 - (\Sigma d_i)^2}{n(n-1)}}$$

$$= \sqrt{\frac{8(100) - (16)^2}{8(7)}} = \sqrt{9.71} = 3.12$$

$$t = \frac{\bar{d} - \mu_d}{S_{\bar{d}}} = \frac{2-0}{3.12/\sqrt{8}} = \frac{2-0}{3.12/2.83} = 1.82$$

6. Reject H_0.

7. The professor-bartender should try to switch his customers to vodka.

In Chapter 4 we saw that knowing the mean of a set of data does not always give us enough information. Thus the concept of a variance or measure of dispersion was introduced to help us describe data more completely. In a similar fashion, we need to augment our knowledge gained thus far about testing hypotheses concerning means with tests of hypotheses concerning variances.

TESTING A SAMPLE VARIANCE VERSUS A POPULATION VARIANCE—THE CHI-SQUARE DISTRIBUTION

To handle a test of hypothesis where we are concerned about whether or not a sample variance is significantly different from a presupposed standard, we need the chi-square (χ^2) distribution. The χ^2 distribution is unlike the normal and the t distribution in that it is skewed (asymmetric). The only parameter of the χ^2 that we need is degrees of freedom. It has been shown that if the variance S^2 is tabulated from a sample of size n and the sample is from a normal population with a variance σ^2, then

SO&B 11-4:
$$\chi^2 = \frac{(n-1)S^2}{\sigma^2}$$

is distributed as a χ^2 distribution with the parameter $n-1$. As long as we know this, we can utilize the probabilities found from the χ^2 distribution. The values of the χ^2 distribution are listed in Table A-5 in the appendix; this table is read the same as the table for the t distribution except that there is no allowance needed for two-tailed tests.

To get a feel for this new distribution, suppose we work through a probability problem. The variance of the width of the pavement on a four-lane expressway is supposed to be 20 inches. Let's assume for the moment that this width is normally distributed. Then what is the probability of getting, in a random sample of 11 places on the road, a variance of 36.614 inches or higher? We calculate

$$\chi^2 = \frac{(n-1)S^2}{2} \quad \text{with } n-1 \text{ df}$$

$$= \frac{(11-1)36.614}{20}$$

$$= 18.307$$

In the χ^2 table, with 10 df, we find that the probability associated with this value is .05. In other words, given a normal distribution with a variance of 20, five times out of a hundred a sample of 11 from this population will have a variance higher than 36.614. Figure 11-1 shows the χ^2 values graphically.

With this introduction to the χ^2 distribution we will put together a test of hypothesis. In Example 2 of the preceding section, we left our Danish prince with the decision to hang around for a while. If you recall, his test of hypothesis showed that the number of slings and arrows he suffered was not significantly greater than his predetermined average. Now he has second thoughts. He ponders that the average number of slings and arrows per day is all right, but life is too much like a yo-yo. Some days he has to suffer a lot and other days

FIGURE 11-1. The χ^2 distribution for $n = 11$.

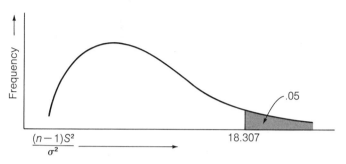

not so much. He can't stand this constant change. The astute reader at this point will exclaim, "Eureka, he's concerned about the variance!" In fact, after careful thought the prince decides that he cannot withstand a daily standard deviation of more than 2.

In the sample of six days the prince examined in the previous section, he found 25, 16, 28, 21, 19, and 22 slings and arrows on the days in his sample. This sample had a calculated variance S^2 of 18.17. The prince now wants to know whether the sample variance S^2 of 18.17 is significantly greater than his predetermined standard for the population variance of 4 (σ^2 is the square of σ; $2^2 = 4$). He assumes that the number of slings and arrows suffered per day follows a normal distribution. Therefore he will follow the seven-step procedure for hypothesis testing, and the testing statistic (step 3) will be

$$\chi^2 = \frac{(n-1)S^2}{\sigma^2}$$

He performs the following test:

1. H_0: $\sigma^2 \leqslant 4$
   H_1: $\sigma^2 > 4$

2. The consequence (suicide) of rejecting a true hypothesis is pretty harsh, so the prince sets $\alpha = .01$.

3. $\chi^2 = \frac{(n-1)S^2}{\sigma^2}$

4. The region of rejection is read from the tables with $\alpha = .01$ and $n - 1 = 6 - 1 = 5$ df. The RR: $\chi^2 > 15.086$.

5. $\chi^2 = \dfrac{(n-1)S^2}{\sigma^2} = \dfrac{(6-1)18.17}{4}$

$\quad = \dfrac{5(18.17)}{4} = \dfrac{90.85}{4} = 22.71$

6. Reject the null hypothesis.

7. End the story—which is a shame, because if he had lived longer he may have made a great idea for a play.

Example 4: A manufacturer of ball bearings is looking at a new production process. It has been shown already that the diameter of the bearings from the new process has an average deviation of zero from the specified diameter. He needs to assure himself, though, that the variance is under control. Obviously, neither his customers nor his quality control people can tolerate a large variance even though the average is adequate. A sample of 21 bearings from the new process shows a variance of .002 inch. The manufacturer wants to know whether this variance is significantly greater than his standard for variance of .0015. He knows that the error from specifications is approximately normally distributed. The test would be this:

1. H_0: $\sigma^2 \leq .0015$
   H_1: $\sigma^2 > .0015$

2. $\alpha = .05$

3. $\chi^2 = \dfrac{(n-1)S^2}{\sigma^2}$ $\quad$ with 20 df

4. RR: $\chi^2 > 31.410$

5. $\chi^2 = \dfrac{(n-1)S^2}{\sigma^2}$

$\quad = \dfrac{(21-1).002}{.0015} = \dfrac{.040}{.0015} = 26.7$

6. Do not reject the H_0.

7. At least continue to consider the new process.

This problem is shown schematically in Figure 11-2.

FIGURE 11-2.   Schematic of Example 4.

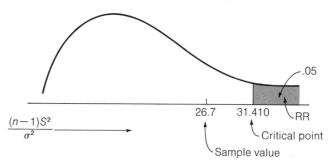

TESTING A SAMPLE VARIANCE VERSUS ANOTHER SAMPLE VARIANCE – THE *F* TEST

Sometimes we have to determine whether two sample variances are significantly different from each other – analogous to a previous situation of determining whether two sample means are significantly different. This situation, under certain conditions, can be handled by the *F* distribution. It can be proved that if random samples of size n_1 and n_2 are taken from two normal populations with the same variance, then the two sample variances S_1^2 and S_2^2, in the ratio

SO&B 11-5: $$F = \frac{S_1^2}{S_2^2}$$

have an *F* distribution with parameters $n_1 - 1$ and $n_2 - 1$.

The tables for selected values of the *F* distribution appear in Table A-6 in the appendix. Notice that these tables are in a slightly different format than tables we have previously used. First of all, each α level is set in a different typeface. Table A-6 includes values for α levels of .01 or .05 only. If you want to find the *F* value associated with an α level of .05 and two samples, each of size 10 and from normal distributions with equal variances, you look across the first row of the 5 percent *F* table to find the degrees of freedom associated with the numerator of SO&B 11-5 ($n_1 - 1 = 10 - 1 = 9$ df). Then you follow that column down to the row associated with the degrees of freedom of the denominator of SO&B 11-5 ($n_2 - 1 = 10 - 1 = 9$ df). The value you find for *F* is 3.18.

In other words, if we take two samples of size 10 from normal distributions with equal variances, 5 percent of the time the ratio (SO&B 11-5) of the sample variances will exceed 3.18. Note that as we take larger and larger samples, it takes less and less difference to be significant. This, of course, is only logical. If we take a very large sample from both populations, then any deviation from a ratio of 1 indicates that the two population variances are not equal.

We are going to contradict a couple of basic points in using the F test for tests of hypotheses concerning two variances. We will do so, however, to simplify the practical usage. The basic tables of F distribution can be manipulated to apply to right-tailed, left-tailed, or two-tailed tests. This is all well and good to the theoretician, but the fact of the matter is that in almost all practical situations we are interested only in the question, "Are the two sample variances significantly different?" We can ignore, then, all the manipulations required by questions of, say, "Is variance 1 less than variance 2?" If you do run into one of these rare situations, merely refer to a standard dry statistics text or the standard dry statistician down the hall.

We can handle the vast majority of problems by forcing the F test to be a right-tailed test. We do this by always making the ratio of the larger sample variance over the smaller sample variance:

SO&B 11-6:
$$F = \frac{S^2 \text{ larger}}{S^2 \text{ smaller}}$$

In this one case, then, we will calculate our sample data *before* we set up the test of hypothesis. When using this technique of always putting the larger variance over the smaller variance, remember that you have in effect doubled the α level of the test. In other words, reading the standard F tables for 5 percent would give you the region of rejection for a 10 percent level of significance.

Suppose we are studying the operating temperatures of two different engines. Their average temperatures are about the same, but we are concerned about whether or not there is a difference in the variances. We also assume, from past data, that these temperatures are at least approximately normally distributed. A sample of 11 (n_1) observations on the first engine shows a variance of 4.5 (S_1^2). A sample of 16 (n_2) observations on the second shows a variance of 3.0 (S_2^2). The test of hypothesis would be this:

1. H_0: $\sigma_1^2 = \sigma_2^2$
   H_1: $\sigma_1^2 \neq \sigma_2^2$

2. $\alpha = .10$

3. $F = S^2$ larger/S^2 smaller (SO&B 11-6) with $n_1 - 1$ df for the numerator and $n_2 - 1$ df for the denominator

4. RR: $F > 2.55$ (from an F table for 5 percent)

5. $F = S^2$ larger/S^2 smaller $= 4.5/3.0 = 1.5$

6. Do not reject H_0

7. The results of the sample are not sufficient to choose the second engine over the first on the basis of variability of temperature.

AUTOPSY ON TESTS CONCERNING VARIANCES

The beginning student is often misled by his intuition concerning variances. Two variances are sometimes erroneously assumed to be different by "eyeballing" them when in fact there are no data to back this assumption up. You must become used to the idea that it takes a heap of difference to be significant because, through the calculation process, a slight difference in one or two of the random observations can cause wide fluctuations in the variance. Remember: You are not working with the regular measurements, as in a standard deviation, but rather with their squares . . . and squares grow surprisingly fast. In short, do not trust your intuition when you are working with variances. Put them to a formal test to avoid being misled.

ANALYSIS OF VARIANCE

Analysis of variance, or experimental design, is a large and important aspect of quantitative analysis. Here we will merely examine its basic concepts and see what analysis of variance can do for you. The basic objective is to make you aware that this type of problem does exist and can be efficiently solved.

We have spent a considerable amount of time now looking at situations and trying to determine whether there is a significant difference between two means. For an example of such a problem, take two supermarkets and compare the average price of a bar of Sudsy Soap.

We observe this price on different days in the two stores and find the results displayed in Table 11.1. We could then attempt to test a hypothesis concerning the difference between prices at the two stores by utilizing a *t* test for independent samples.

TABLE 11-1. Price Per Bar of Sudsy Soap

Observation	Store 1	Store 2
1	.13	.23
2	.15	.19
3	.11	.19
4	.15	.23
5	.16	

We might be concerned, though, with five stores instead of two. In this case, we would observe the price of Sudsy Soap in each of the stores on different days. The results could be as shown in Table 11-2.

TABLE 11-2. Price Per Bar of Sudsy Soap

Observation	Store 1	2	3	4	5
1	.13	.23	.24	.18	.21
2	.15	.19	.22	.17	.22
3	.11	.19	.28	.21	.17
4	.15	.23	.26	.15	.17
5	.16		.24		.20
6			.27		.21
Sum	.70	.84	1.51	.71	1.18 = 4.94
Average	.140	.210	.252	.178	.197 = .198

Here we are concerned about whether or not the prices are different at the five different stores. Thus we want to test the hypothesis that

$$H_0: \mu_1 = \mu_2 = \mu_3 = \mu_4 = \mu_5$$

or all means are equal, against the hypothesis that

H_1: All means are not equal

Analysis of variance provides a way of accomplishing this task. In other words, it is a body of statistical techniques that helps us to make inferences about more than two population means.

This example is the most basic sort of analysis-of-variance problem. With analysis of variance we are comparing variances to make inferences about means. In this example we will look at the variance of the *means* of the five stores and in concept compare this variance to the variance of the prices in total. Roughly, then, if the variance of the means is more than can be accounted for by the variance of prices in total, the means are significantly different.

One bit of additional notation, referred to as *dot notation*, must be introduced here. Table 11-2 represents a two-dimensional array of numbers. The observation x_{ij} is the price in the ith row and jth column. For example, x_{23} is 22 cents. The notation $x_{.j}$ indicates that the x's have already been added together for all i's (or a column has been added). The notation $\bar{x}_{.j}$ would be the mean of the jth column, and $\bar{x}_{..}$ would be the grand total (all x's added over all i's and j's). The term n_j refers to the number of observations in the jth column.

By referring to Table 11-2 you should be able to verify:

$$x_{11} = .13$$
$$x_{34} = .21$$
$$\bar{x}_{.1} = .140$$
$$\bar{x}_{.2} = .210$$
$$x_{.1} = .70$$
$$x_{.2} = .84$$
$$x_{..} = 4.94$$
$$\bar{x}_{..} = .198$$
$$n_1 = 5$$

The testing statistic we will use for this basic example of analysis of variance is

SO&B 11-7:
$$F = \frac{\sum_{j=1}^{r} n_j(\bar{x}_{.j} - \bar{x}_{..})^2 \big/ (r-1)}{\sum_{j=1}^{r} \sum_{i=1}^{n_j}(x_{ij} - \bar{x}_{.j})^2 \big/ (N-r)}$$

where r is the number of columns and N is the total number of observations. SO&B 11-7 will follow an F distribution if the following two assumptions are met:

1.   The samples are random and come from normal distributions.

2.   The means of the distributions are equal and the variances of the distribution are equal.

The F distributions will have the parameters $(r-1)$ df for the numerator and $(N-r)$ df for the denominator. This ratio boils down to a much simpler computational form, the same as

$$S = \sqrt{\frac{\Sigma(x - \bar{x})^2}{n-1}}$$

reduced to

$$S = \sqrt{\frac{n\Sigma x^2 - (\Sigma x)^2}{n(n-1)}}$$

for computational ease. However, we will use the definitional form for our purposes here.

In our example, the numerator of the F ratio is $5(.14 - .198)^2$ $+ 4(.21 - .198)^2 + 6(.252 - .198)^2 + 4(.178 - .198)^2 + 6(.197 - .198)^2$ divided by $r-1$ or 4. This numerator thus equals $3.6364/4 = .9091$. The denominator of the F ratio is $(.13 - .140)^2 + (.15 - .140)^2$ $+ (.11 - .140)^2 + (.15 - .140)^2 + (.16 - .140)^2 + (.23 - .21)^2 + \ldots$ $+ (.21 - .197)^2$ divided by $N - r$ or 20. (The "$+ \ldots +$" indicates that data following an identical format have been left out for brevity.) This denominator thus equals $.9892/20 = .0495$.

Our computed F ratio, then, is

$$F = \frac{.9091}{.0495}$$

$$= 18.37$$

Putting this information into the seven-step test of hypothesis, we have:

1. H_0: $\mu_1 = \mu_2 = \mu_3 = \mu_4 = \mu_5$ (or all means are equal)
    H_1: All means are not equal

2. $\alpha = .05$

3. $F = \dfrac{\sum\limits_{j=1}^{r} n_j(x_{.j} - x_{..})^2 \Big/ (r-1)}{\sum\limits_{j=1}^{r} \sum\limits_{i=1}^{n_j} (x_{ij} - x_{.j}) \Big/ (N-r)}$ with $(r-1)$ and $(N-r)$ df

4. $F > 2.87$ ($F_{.05}$ with 4 and 20 df)

5. $F = .9091/.0495$

    $= 18.37$ (from our previous calculation)

6. Reject H_0.

7. Find the reason for the price variation in different stores.

If you haven't choked a few times during this discussion of analysis of variance, then, as the saying goes, you probably don't understand the seriousness of the situation. However, remember that this is merely an introduction to the topic; it is presented here to inform you this technique of analysis exists. This type of example would gain in complexity if, for instance, we were also concerned about whether stores 1 and 2, which are in one area of the city, have significantly different prices than stores 3, 4, and 5 in another area of the city. There are designs and methods of analysis of variance to handle almost any subdivision or multiple subdivision of the data. In fact, analysis of variance is an entire field of study in itself.

AUTOPSY

In this chapter we have looked at a number of hypothesis-testing situations. The mechanics of working through each one become relatively painless after a little practice and should be considered a cut-and-dried part of statistics. The real crux of the matter in working situations is choosing which test to use in a certain situation. Also involved in the "art" of statistics is choosing an appropriate level of significance (α). In business we often use a higher α level than is commonly used in research statistics.

SUMMARY

The following table summarizes the situations in which you would use a certain testing statistic. The choice of a certain test of hypothesis (which determines the testing statistic) depends on the question that needs answering (the hypothesis) and the assumptions or conditions of the situation.

Question Concerns	Conditions	Testing Statistic
Sample mean vs. population mean	σ is known or $n > 30$	$z = \dfrac{\bar{x} - \mu}{\sigma_{\bar{x}}}$
Sample proportion vs. population proportion	nP and $nQ > 5$	$z = \dfrac{p - P}{\sqrt{PQ/n}}$ or $z = \dfrac{x - nP}{\sqrt{nPQ}}$
Sample mean vs. population mean	σ is unknown and $n \leqslant 30$; x's are normally distributed	$t = \dfrac{\bar{x} - \mu}{S_{\bar{x}}}$
Difference between two sample means	Independent samples; samples from normal distributions with equal variances	$t = \dfrac{\bar{x}_1 - \bar{x}_2}{S_{\bar{x}_1 - \bar{x}_2}}$

$$\text{where } S_{\bar{x}_1 - \bar{x}_2} = \sqrt{\frac{(n_1 - 1)S_1{}^2 + (n_2 - 1)S_2{}^2}{n_1 + n_2 - 2}} \times \sqrt{\frac{n_1 + n_2}{n_1 \cdot n_2}}$$

Difference between two sample means	Dependent samples	$t = \dfrac{\bar{d}}{S_{\bar{d}}}$
Sample variance vs. population variance	Sampling from normal distribution	$\chi^2 = \dfrac{(n - 1)S^2}{\sigma^2}$
Difference between two sample variances	Sampling from normal distribution	$F = \dfrac{S^2 \text{ larger}}{S^2 \text{ smaller}}$
Difference between more than two sample means	Samples from normal distributions with equal means and equal variances	Analysis of variance

PROBLEMS

1.   A manufacturer of dry cereal is producing 20-ounce packages of his product. The automated packaging device is checked periodically to see whether it is actually putting 20 ounces in each package. The weights of the finished packages are thought to be normally distributed. A sample of nine packages shows that $\bar{x} = 19.8$ and $S^2 = .25$.

   a.   Test the appropriate hypothesis at a 5 percent level of significance.

   b.   How does this problem differ from Problem 2 in Chapter 10?

2.   A toy distributor is buying batteries for electric toys. His supplier guarantees that his brand of batteries will last 6.5 hours. After receiving some customer complaints, the toy distributor randomly selects six batteries from his stock and tests them for life expectancy. The results are as follows:

Battery	Lifetime (hr)
1	7.0
2	6.5
3	5.9
4	6.6
5	6.2
6	6.1

Assuming that the lifetime of these batteries is normally distributed, test the hypothesis at a 10 percent level of significance.

3.   This problem has been developed to emphasize that statistics is not so cut and dried as some people would have you believe. The City Dump Mining Company is exploring the possibility of reclaiming minerals from a metropolitan garbage dump. One of the items of concern to them is the pounds of aluminum per ton of waste material. They estimate that they need at least 120 pounds per ton (ppt) to make their operation profitable. A sample of ten test pits shows an average of 128 ppt with a standard deviation of 20 ppt. They want to conduct a test of hypothesis.

a. Conduct the test of hypothesis at a 5 percent level of significance with the null hypothesis being $\mu \leqslant 120$.

b. Conduct the test of hypothesis at a 5 percent level of significance with the null hypothesis being $\mu \geqslant 120$.

c. Compare parts (a) and (b) and discuss their implied management decisions.

d. What circumstances could make you choose the hypothesis from (a)?

e. What circumstances could make you choose the hypothesis from (b)?

4. Weird Willy, the statistical consultant, catches a commuter bus home every night. There is only one bus and if he misses it he has to take a taxi at an exorbitant cost. He also is very busy, so he times himself closely. The bus doesn't arrive at the stop at exactly the same time every day, though. Willy figures that the bus gets to the stop at 5:10 P.M. on the average, and he will leave his office at precisely the right time to catch the bus at that time. He times the bus on five days with the following results:

Day	Bus Arrival
1	5:12
2	5:07
3	5:09
4	5:07
5	5:10

a. Test the hypothesis that $\mu = 5{:}10$ at a 1 percent level of significance.

b. Why would this analysis be totally invalid for Willy?

5. A manufacturing plant is concerned about the high number of serious grievances received. A consultant team sells them on the idea of "sensitivity" training sessions for the foremen. These training sessions will be costly, so they want to analyze the results to see whether they are effective. They keep records on the number of complaints received by the employees under eight of their foremen for one month before the sessions and one month after. The results are as follows:

Foreman	No. of Complaints Before	After
1	14	9
2	22	24
3	10	6
4	15	17
5	16	13
6	21	14
7	20	20
8	15	16

a.   Is this a dependent or an independent sample?

b.   Test the appropriate hypothesis at a 10 percent level of significance.

6.   A mining equipment company has the chance to purchase a new line of drill bits. The new bits cost the same amount but are supposed to last longer. The purchasing agent acquires a dozen of the new bits to test them out. He asks one of his customers to keep a record on the lifetime of six of the new bits and eight old bits used in comparable rock. In return the customer is given a special discount. The customer supplies the following data:

Lifetime of Old Bits (hr)	Lifetime of New Bits (hr)
17	16
17.5	18.1
16	17.6
17.4	16.9
18	17.9
16.9	18.4
16.5	
17	

The purchasing agent calculates the following:

	New	Old
Mean	17.5	17.0
SD	.88	.62

Test the appropriate hypothesis at $\alpha = .10$.

7. A candy store has designed a one-week promotion to increase the dollar sales per customer. The manager would like to know whether the promotion was successful, so he takes a random sample of ten sales slips before the promotion and a sample of ten sales slips during the promotion. An analysis of these sales slips shows the following:

Before	During
$n_b = 10$	$n_d = 10$
$\bar{x}_b = \$2.14$	$\bar{x}_d = \$2.48$
$S_b^2 = \$0.25$	$S_d^2 = \$0.39$

Test the appropriate hypothesis at $\alpha = .10$.

8. The health inspector for an urban area is checking the fly content of pea soup. He is concerned about whether the fly content of pea soup in the central area is higher than that in the suburban area. On a certain day he and his staff check 12 central area restaurants and find that the average number of flies per bowl of soup is 2.4 with a standard deviation of .6. A check of 14 suburban restaurants shows an average of 1.7 flies per bowl with a standard deviation of .7. The inspector is willing to assume that the samples come from normal populations with equal variances.

    a. Are the samples dependent or independent?

    b. Test the appropriate hypothesis at a 5 percent level of significance.

9. A taxicab company is approached by a salesman who claims to have a device that will reduce gas consumption in a car. It is inex-

pensive and easy to install. The salesman will allow the taxicab company to test six devices to see if they really work. He has the gas mileage records for six of the taxicabs and their regular drivers. He installs the test devices in these cars and, with the same regular drivers for each, gets the following results:

Car	Gas Mileage	
	With Device	Without Device
1	19.7	19.1
2	20.0	20.1
3	19.5	19.8
4	18.9	18.2
5	19.3	19.4
6	19.7	19.3

a. Why is it important to use the same cars and drivers?

b. What other variables should be watched?

c. Test the appropriate hypothesis at $\alpha = .05$.

10. A machine is supposed to be adjusted to mold 3-ounce lead weights. A sample of 12 weights has an average weight of 2.940 ounces and a standard deviation of .10 ounce. Is this sample result consistent with the machine being in adjustment? Assume that the weights are normally distributed and test the appropriate hypothesis at a 5 percent level of significance.

11. A buyer of the lead weights from Problem 10 uses them in a delicate balancing operation. He therefore wants them to be consistent in their weight. In fact, he will not accept a variance in the weights of more than .005 ounce. Test the appropriate hypothesis at a 5 percent level of significance.

12. A manufacturing plant has thousands of ceiling light fixtures. Because of the height of the ceiling and the special climbing equipment required, the plant manager has found that it is cheaper for him to have all the light bulbs changed at one time rather than replacing them individually. The brand of bulb he now uses has an average life of 1050 hours and a known standard deviation of 50 hours. He has also noted that the life expectancy of the bulbs is approximately normally distributed.

Now a new and slightly cheaper brand of light bulbs is being considered. A sample of 26 new bulbs shows an average life of 1050 hours and a standard deviation of 60 hours. The variance of the new bulbs concerns the plant manager because if a few bulbs burn out very quickly he will either have less light or have to change all the bulbs more often. Test the hypothesis for the plant manager that the new bulbs have a variance less than or equal to the old standard $(H_0: \sigma^2 \leq 50)$.

13.   In Problem 7, test the hypothesis that the variances of the sales amounts are the same before and after the promotion.

14.   In problem 6, test the hypothesis that the variances of the two types of bits are the same.

15.   An executive of a large company has consented to send 15 of his middle management personnel to a speed-reading course next week. For future reference he needs to know whether there will be a significant improvement in reading speed. If he is to set up the analysis in the most efficient manner and then conduct a test of hypothesis after he receives the appropriate data, what should be the testing statistic of this best test?

16.   Iron Mike is running for public office and is soliciting campaign donations. He has two brochures he can use in these solicitations and he wants to find out which is more effective. He randomly mails 20 of one kind to one group and 20 of the other kind to another group. One brochure shows pictures of him running hippies, students, college professors, and other communists out of town. The other brochure shows pictures of him petting stray dogs, kissing babies, and attending Lion's Club meetings. He will measure the effectiveness of each brochure by tabulating the average donation received. In a test of hypothesis concerning the two brochures, what could be the testing statistic of the most appropriate test?

17.   A product can be sold through four different types of outlet. In a one-week test the product is put in a sample of each of the four different types and the sales results are recorded. Management, of course, is interested in knowing whether the sales level is the same for all four types of outlet. What technique could be applied to analyzing the sample results?

18.   A large company has been faced with a high turnover rate among its secretarial and clerical staff. Some of these people have been hired through the company's own personnel department and

others have been hired through the Smiley Employment Agency. The personnel manager would now like to know whether the average length of employment is different for the two sources of employees. Briefly describe the best testing procedure he could use and state the appropriate testing statistic.

19. In a manufacturing plant there are three employees making the same electronic device. The plant manager would like to know whether there is a significant difference in the number of defects produced by each of the three employees. A random sample of five 8-hour shifts shows the following number of defects:

Shift	Employee A	Employee B	Employee C
1	12	15	7
2	10	8	14
3	14	15	10
4	9	12	8
5	12	12	9

We are willing to assume that the number of defects is normally distributed. Do an analysis of variance to answer the plant manager's concern at a 5 percent level of significance.

SELECTED REFERENCES

Dixon, W. J. and F. J. Massey. *Introduction to Statistical Analysis.* Third edition. New York: McGraw-Hill, 1969.

Guenther, William C. *Concepts of Statistical Inference.* Second edition. New York: McGraw-Hill, 1973. Especially Chapters 5, 7.

Miller, Irwin and John E. Freund. *Probability and Statistics for Engineers.* Englewood Cliffs, N.J.: Prentice-Hall, 1965. Especially Chapters 9, 13, 15.

Tanur, Judith M., Frederick Mosteller, William H. Kruskal, Richard F. Link, Richard S. Pieters, and Gerald R. Rising. *Statistics: A Guide to the Unknown.* San Francisco: Holden-Day, 1972.

Yamane, Taro. *Statistics: An Introductory Analysis.* Third edition. New York: Harper & Row, 1973. Especially Chapters 18, 21, 22.

CHAPTER 12
ANALYSIS OF FREQUENCIES (χ^2)

OBJECTIVES: After studying this chapter and working the problems, you should be able to:

Demonstrate when and how to make chi-square calculations to see whether a counted sample fits a presupposed proportion.

Demonstrate when and how to use contingency tables to determine independence of characteristics.

Demonstrate when and how to use goodness-of-fit tests to see whether a certain distribution can be assumed.

IN THIS CHAPTER we will work with a certain kind of test of hypothesis that can be classified as a *nonparametric test*. A nonparametric or *distribution-free* test is one that does not require knowledge of any parameter of the population. For instance, in working with the normal distribution as in previous chapters we have to know the parameters μ and σ. In other tests we conducted, we had to assume that the data came from a normal distribution. A nonparametric test does not need these parameters and assumptions.

ANALYSIS-OF-FREQUENCY TESTS

One such nonparametric test is the χ^2 analysis of frequencies. This test can be used when we are counting the number of observations that fit a certain category and want to test certain characteristics of our counted observations. We will look at three different types of χ^2 analysis-of-frequency tests:

1. Testing data to see whether they differ significantly from a presupposed proportion. For example: A shirt manufacturer may be interested in knowing whether the three different types of shirts he makes still sell in the ratio of 5:3:2.

2. Testing to determine whether two different qualities or characteristics of a population are independent or *contingent* upon one another. For example: A marketing manager may be interested in knowing whether the age of a customer and the color of package preferred are contingent upon one another.

3. Testing to determine whether the data come from a certain probability distribution or a *goodness-of-fit test.* For example: A production manager may know that the analysis of the weights of his products would be much easier if he could safely assume that the weights follow a normal distribution.

We will look at each of these three types of test. The testing statistic we will be using is

$$\chi^2 = \sum_{i=1}^{k} \frac{(O_i - E_i)^2}{E_i}$$

with appropriate degrees of freedom. We will also be using the tabled values of χ^2 in Table A-5 in the appendix. In the preceding equation the O_i represents the number of sample observations that were counted in the ith category and E_i represents the number of observations we would "expect" to be there. Now let's look at each of the three types in detail. Remember: In all these situations we are counting frequencies.

TESTING SAMPLE FREQUENCIES AGAINST A PRESUPPOSED PROPORTION

Imagine that you are the sales manager for a firm producing men's shirts. You manufacture shirts of three different quality levels that sell for three different prices. Being imaginative, you have labeled these three types as neat, sharp, and supersharp. The production people now need to know what percentage of production to put into each type of shirt for the coming sales period so that they can make the proper purchases and machine adjustments. Recently your sales have been running 20 percent in the neat line, 50 percent sharp, and 30 percent supersharp. You think there may be a change in that distribution because of the effects of inflation. A sample of retail shirt sales from stores that handle your product show the following:

Type of Shirt	Dozens Sold
Neat	23
Sharp	55
Supersharp	22
	100

The question now is this: Are these sample results significantly different from the 20:50:30 split we had in the past?

We begin by calculating the number we would have "expected" to find in each category if the 20:50:30 split still held. We do this in our example by noting that there was a total of 100 observations. Therefore if 20 percent of the sales were still "neat," we would expect 20 percent of 100, or 20, observations in that category. Similarly, we would expect sales of 50 sharp and 30 supersharp shirts. We will call the actual observations O_i ($O_1 = 23, O_2 = 55, O_3 = 22$) and the expected values E_i ($E_1 = 20, E_2 = 50, E_3 = 30$).

We know that the value of

SO&B 12-1: $$\chi^2 = \sum_{i=1}^{k} \frac{(O_i - E_i)^2}{E_i} \quad \text{with } k - 1 \text{ df}$$

is approximately distributed as a χ^2, where k is the number of categories (not the number of observations, but the number of categories). Hence SO&B 12-1 will be our testing statistic in this situation.

Notice in SO&B 12-1 that the further apart the expected and observed values are, the larger the value of χ^2 will be. It is logical, therefore, that if the χ^2 value is sufficiently large, we will *reject* the idea that the observed values are comfortably close to our expected values. Putting this into a formal test of hypothesis we have the following:

1. H_0: Sales are distributed 20:50:30
   H_1: Sales are not distributed 20:50:30

2. $\alpha = .10$

3. $\chi^2 = \sum_{i=1}^{k} \frac{(O_i - E_i)^2}{E_i}$ with $k - 1 = 3 - 1 = 2$ df
   (SO&B 12-1)

4. RR: $\chi^2 > 4.605$ (from Table A-5 note that the region of rejection will always be right-tailed in this type of test)

5.

Type of Shirt	Presupposed Proportions	Observed Values (O_i)	Expected Values (E_i)	$(O_i - E_i)^2$	$\frac{(O_i - E_i)^2}{E_i}$
Neat	20%	23	20	$3^2 = 9$	.45
Sharp	50%	55	50	$5^2 = 25$	.50
Supersharp	30%	22	30	$-8^2 = 64$	2.13
		100	100		$\chi^2 = 3.08$

6. Because 3.08 does not fall in the region of rejection, we would not reject H_0.

7. Tell the production people not to change anything yet.

For further drill in this procedure, let's go through another example. Our old friend Mabel Metcher has taken up playing the dice table in the back room of Bud's Bookstore. (Bud is a state senator who continually gets elected by promising a cleanup of illegal gambling in the city.) Mabel has been losing rather consistently and is beginning to suspect that the dice may not be fair, so she starts to keep a record of what comes up on one of the dice. At the end of the evening she has this record:

Die Face	Times Appeared
1	27
2	11
3	14
4	21
5	19
6	28
	120

In past problems concerning dice we have been limited to the value of one face—say the proportion of the time a 6 appears. With the χ^2 as a new tool, we can now take the whole thing at once. First we find our expected values. If the dice are fair, we would expect each of the six faces to appear one-sixth of the time. There were 120 throws tabulated, so our expected value for each face would be one-sixth of 120, or 20 times. We now make our formal test of hypotheses:

1. H_0: Die is fair
    H_1: Die is unfair

or

   H_0: Faces are distributed 1:1:1:1:1:1
   H_1: Faces are not distributed 1:1:1:1:1:1

2. $\alpha = .05$

3. $\chi^2 = \sum_{i=1}^{k} \frac{(O_i - E_i)^2}{E_i}$ with $k - 1 = 6 - 1 = 5$ df

4. RR: $\chi^2 > 11.070$

5.

Die Face	Observed Frequency (O_i)	Expected Frequency (E_i)
1	27	20
2	12	20
3	14	20
4	20	20
5	19	20
6	28	20
	120*	120*

* If these two sums are not equal, you have made an arithmetic error.

$$\chi^2 = \frac{(27-20)^2}{20} + \frac{(11-20)^2}{20} + \frac{(14-20)^2}{20} + \frac{(21-20)^2}{20}$$

$$+ \frac{(19-20)^2}{20} + \frac{(28-20)^2}{20}$$

$$= \frac{232}{20} = 11.6$$

6. Reject H_0.

7. Mabel should yell "Cheats!" . . . after she is safely out the door.

The χ^2 test as used here does not tell us which category has caused the transgression if H_0 is rejected; however, this can usually be determined by eyeballing the data.

In the two preceding examples the expected values came out to be integers (whole numbers) although this does not have to be the case. If the expected values turn out uneven they should be carried out to two decimals. You say that an observed counted frequency cannot be a fraction? This is true, but remember that these are theoretically expected values. To round them to a whole number can introduce serious errors.

CHI SQUARE IN CONTINGENCY TABLES

In the section concerning statistical description, we talked about cross-classification—a procedure whereby we sort observations according to two or more characteristics (such as sex versus income) within one chart. The χ^2 distribution gives us a means of expanding our analysis of a cross-classification. Sometimes we are concerned about whether or not two qualities or characteristics are interrelated (dependent). We may want to know whether the color of packaging is interrelated with the age of the purchaser or whether family income is interrelated with the number of movies attended or whether the number of defects produced is interrelated with the day of the week. In these situations we can use the *contingency table test*.

Let's take one of these situations to illustrate the procedure and use of a contingency table test. Suppose that we have sent out a survey

TABLE 12-1. Number of
Questionnaire Responses

| Color of Package | Age of Respondent | | |
Preferred	21–30	31–40	41–50
Red	20	12	8
Yellow	7	3	10
White	13	15	12

team to get information about packaging for a new product. The data from the survey are summarized and, among other things, Table 12-1 is produced from the responses.

Management now wants to know whether different age groups prefer different colors. We can translate that problem into a statistical question: Is the characteristic "age of respondent" independent of the characteristic "color of package preferred"? We will use the same testing statistic again, namely,

SO&B 12-2: $\quad \chi^2 = \sum \dfrac{(O_i - E_i)^2}{E_i} \quad$ with $(r - 1) \cdot (c - 1)$ df

This is the same as SO&B 12-1 except that this time the degrees of freedom are the number of rows (r) minus 1 times the number of columns (c) minus 1 — or $(r - 1) \cdot (c - 1)$. In our table we have three rows and three columns so that df $= (3 - 1) \cdot (3 - 1) = 4$.

Each cell represents an observed value (O_i). We need to develop the expected value (E_i) for each cell and can do this by applying a bit of logic. First we get the totals for each row and column as shown in Table 12-2.

TABLE 12-2. Number of Questionnaire
Responses

Color of Package Preferred	Age of Respondent			
	21–30	31–40	41–50	Total
Red	20	12	8	40
Yellow	7	3	10	20
White	13	15	12	40
	40	30	30	100

Now that we've totaled each row and each column, it's time to apply the promised logic. There were 100 respondents all together. Of these 100 respondents, 40 preferred the red package. Now if the package color and the age are independent, we would expect 40 one-hundredths (.40) of the 21 to 30 age bracket to prefer red — or .40 · 40 (number of 21 to 30 year olds) = 16. Sixteen, then, is the number of respondents between 21 and 30 years old that we would *expect* to prefer red if the two qualities are independent.

Don't quit now! We'll go over the procedure once more. See if you can follow it through step by step. Again, 40 percent of the total respondents preferred red (40 out of 100). Therefore if age and package color are independent, we would expect 40 percent of the 31 to 40 year olds to prefer red—or $.40 \cdot 30$ (number of 31 to 40 year olds) = 12. We would expect *twelve* respondents to be between 31 and 40 *and* prefer red packages *if* the two characteristics are truly independent. Algebraically the two calculations we just did would appear as follows:

$$\text{Expected no. of} \atop \text{21–30 } and \text{ red} = \frac{\text{no. of reds}}{\text{total observations}} \cdot \text{no. of 21–30's}$$

$$= \frac{40}{100} \cdot 40$$

$$= 16$$

$$\text{Expected no. of} \atop \text{31–40 } and \text{ red} = \frac{\text{no. of reds}}{\text{total observations}} \cdot \text{no. of 31–40's}$$

$$= \frac{40}{100} \cdot 30$$

$$= 12$$

Let's calculate two more "expected" values now. This may seem tedious the first time through, but once you catch on to the pattern it goes down easily. Next we'll take on the expected number of 21 to 30 year olds *and* yellow preference. Twenty of the total 100 respondents preferred yellow. Therefore we would expect 20 percent of the 21 to 30 year olds to prefer yellow—or $.20 \cdot 40 = 8$. Similarly, we would expect 20 percent of the 31 to 40 year olds to prefer yellow. Therefore $.20 \cdot 30 = 6$, as shown in the first two columns and first two rows of Table 12-3.

As for the rest of the "expected values," remember that we said this problem had 4 degrees of freedom. We have now calculated four expected values. The remainder are "locked in" because the row and column totals are already set and cannot be altered. A total of 40 respondents prefer red. We "expect" 16 of these to be 21 to 30 year olds and we expect 12 of them to be 31 to 40 year olds. By subtraction, then, $40 - 16 - 12 = 12$; thus the remaining 12 of them would be expected to be 41 to 50 year olds. Each of the remaining four cells in the third row and the third column can be filled in with an expected value as shown in Table 12-3.

Note that the five expected values found by subtraction can also be determined by the same procedure as the original four. The result is identical; it is merely easier to do the subtraction. Note also that the expected values are found by utilizing only the "total" column and the "total" row. The actual observed number of respondents in each individual cell does not enter into the calculations.

TABLE 12-3. Expected Number of Respondents
(If Independent)

Color of Package Preferred	Age of Respondent			
	21–30	31–40	41–50	Total
Red	$\dfrac{40}{100} \times 40 = 16$	$\dfrac{40}{100} \times 30 = 12$	$40 - 16 - 12 = 12$	40
Yellow	$\dfrac{20}{100} \times 40 = 8$	$\dfrac{20}{100} \times 30 = 6$	$20 - 8 - 6 = 6$	20
White	$40 - 16 - 8 = \underline{16} \\ 40$	$30 - 12 - 6 = \underline{12} \\ 30$	$30 - 12 - 6 = \underline{12} \\ 30$	40

Now, let's see, "where are we?" We started out by trying to determine whether the age of the respondent is independent of the color of package preferred. We can analyze this problem by using the χ^2 distribution in a contingency table test. To do this test we must know the "expected" number of observations in each cell. We have just gone through calculating these expected values. Now we should put this all in the form of a formal test of hypothesis:

1. H_0: Age is independent of color preference (always hypothesize *in*dependence in a contingency table)
   H_1: Age is not independent of color preference

2. $\alpha = .05$

3. $\chi^2 = \sum \dfrac{(O_i - E_i)^2}{E_i}$ with $(r-1) \cdot (c-1)$ df (SO&B 12-2)

4. $\chi^2 > 9.49$ ($\alpha = .05$ and 4 df)

5. Taking the observed values O_i from Table 12-2 and the expected values E_i from Table 12-3, we have

$$\chi^2 = \sum \dfrac{(O_i - E_i)^2}{E_i}$$

$$= \frac{(20 - 16)^2}{16} + \frac{(12 - 12)^2}{12} + \frac{(8 - 12)^2}{12} + \frac{(7 - 8)^2}{8}$$

$$+ \frac{(3 - 6)^2}{6} + \frac{(10 - 6)^2}{6} + \frac{(13 - 16)^2}{16} + \frac{(15 - 12)^2}{12}$$

$$+ \frac{(12 - 12)^2}{12}$$

$$= 7.9375$$

6.   This is not in the region of rejections; hence we do not reject H_0.

7.   Make no marketing outlays that try to distinguish colors for different age groups.

Notice that in step 7 we translated our analysis back into the original management question.

Example 1:   For those of you who are strong of heart, let's move to a true example in which the names and numbers have been modified to protect the innocent. The problem, being real, is fairly complex in thought but not in actual calculation. The situation is this. A company is trying to "salvage" their sales of stereo units (they sell through retail outlets). The market research department has just completed extensive interviews with the owners of the retail outlets. The retail outlets can be readily classified into those that sell primarily home console stereos, those that sell primarily portable stereos, and those that sell primarily automobile stereo units. The company is now planning an advertising campaign that can be put into national media (magazines, for instance), into local media (newspapers, for instance), into point-of-sale materials (posters, handouts, and the like), or into some combination of these. Of course, a combination will substantially dilute or split the advertising budget.

The question from management is this: "Can we put all our advertising budget in one medium, or do the three different types of dealers (outlets) require different types of advertising?" This basic management question can be translated into statisticalese as follows: "Are the two characteristics, type of dealer and type of advertising preferred, independent?" The frequency count from the interviewer is shown in Table 12-4.

TABLE 12-4.   Observed Number of Dealers

Type of Dealer	Type of Advertising Preferred			Total
	National	Local	POS*	
Console	81	20	14	115
Portable	72	15	23	110
Car	68	32	25	125
	221	67	62	350

* POS = point of sale.

Setting this up in our formal test of hypothesis we have the following:

1. H_0: Type of dealer is independent of type of advertising preferred

    H_1: Type of dealer is not independent of type of advertising used

2. $\alpha = .10$

3. $\chi^2 = \sum \frac{(O_i - E_i)^2}{E_i}$ with $(r-1) \cdot (c-1) = (2) \cdot (2) = 4$ df

4. RR: $\chi^2 > 7.779$

5. The expected values are as shown in Table 12-5. Using the observed values from Table 12-4 and the corresponding expected values from Table 12-5, we calculate:

TABLE 12-5.   Expected Number of Dealers

Type of Dealer	Type of Advertising Preferred			Total
	National	Local	POS	
Console	$\frac{115}{350} \times 221$ $= 72.61$	$\frac{115}{350} \times 67$ $= 22.01$	$115 - 72.61$ $-22.01 = 20.37$	115
Portable	$\frac{110}{350} \times 221$ $= 69.46$	$\frac{110}{350} \times 67$ $= 21.06$	$110 - 69.46$ $-21.06 = 19.49$	110
Car	$221 - 72.61$ $-69.46 = 78.93$	$67 - 22.01$ $-21.06 = 23.93$	$125 - 78.93$ $-23.93 = 22.14$	125
Total	221	67	62	350

$$\chi^2 = \frac{(81 - 72.61)^2}{72.61} + \frac{(20 - 22.01)^2}{22.01} + \frac{(14 - 20.32)^2}{20.32}$$

$$+ \frac{(72 - 69.46)^2}{69.46} + \frac{(15 - 21.06)^2}{21.06} + \frac{(23 - 19.49)^2}{19.49}$$

$$+ \frac{(68 - 78.93)^2}{78.93} + \frac{(32 - 23.93)^2}{23.93} + \frac{(25 - 22.14)^2}{22.14}$$

$$= 10.213$$

6.   Reject H_0.

7.   Direct the advertising to the type of dealer.

Note that in this example the expected values were carried to two decimal places. Do not round your expected values to the nearest whole number. Let's move now to the third type of analysis-of-frequency test.

THE χ^2 GOODNESS-OF-FIT TEST

We have emphasized in preceding chapters that to apply some of our tests of hypotheses we must assume normality of the data. Now experience has shown that the tests will remain valid with considerable variation from this assumption. This is particularly true for tests concerning means; tests concerning variances are more sensitive to deviations from normality. We need to know whether or not the assumption of normality is valid—either to satisfy ourselves or some cynical antagonist. The χ^2 goodness-of-fit test is a procedure for testing this assumption. It is in fact a procedure for determining whether the data follow *any* particular distribution.

   In general, we can break our data into groups (the number of groups is largely a subjective decision made by the analyst) and then compare the frequency observed in each group to the frequency we would expect if the assumed distribution were true. Our testing statistic is

SO&B 12-3: $\chi^2 = \sum \frac{(O_i - E_i)^2}{E_i}$ with $k - w - 1$ df

where k is still the number of groups or categories and w is the number of estimated parameters. The value of w will vary with the type of distribution in which we are interested.

   Let's look at a test of normality first. Suppose we are packaging

TABLE 12-6. Analysis of Bag Weights

Weight (lb.)	(1) Observed Frequency	(2) Normal Probability	(3) Expected Frequency
98 or less	9	.0228	6.84
98.01–99	36	.1359	40.77
99.01–100	107	.3413	102.39
100.01–101	98	.3413	102.39
101.01–102	42	.1359	40.77
More than 102	8	.0228	6.84
	300	1.0000	300.00

monkey food in 100-pound bags for zoos. The contents of the bags is not, of course, always exactly 100 pounds. We want to know, though, for future analysis, whether the weights are normally distributed. We take a sample of 300 bags and record the results as shown in column 1 of Table 12-6. The sample data have a mean of 100 and a standard deviation of 1. The question is this: Can it be assumed that these data came from a normal population?

Column 2 has been developed from Table A-3 in the appendix. If the data are normally distributed with $\mu = 100$ and $\sigma = 1$, then 98 pounds is two standard deviations below the mean. A quick check of the normal distribution tables will show you that .0228 of the data of a normal distribution will be more than two standard deviations below the mean. Similarly, .1359 of the data will be between 1 and 2 σ's away (or 99 to 98 pounds), and the rest of column 2 is found in the same manner. The expected frequency, column 3, is found by multiplying the element of column 2 by the total sample size (the total of column 1).

Our formal hypothesis test would be as follows:

1. H_0: Population is normally distributed
    H_1: Population is not normally distributed

2. $\alpha = .05$

3. $\chi^2 = \sum \dfrac{(O_i - E_i)^2}{E_i}$ with $k - w - 1$ df (SO&B 12-3)

For tests of normality we estimated two parameters, μ and σ, by $\bar{x}$ and S. Therefore $w = 2$ and $k - w - 1 = 6 - 2 - 1 = 3$ df

4. $\chi^2 \geq 7.81$

5. From Table 12-6 we compute:

$$\chi^2 = \frac{(9-6.84)^2}{6.84} + \frac{(36-40.77)^2}{40.77} + \frac{(107-102.39)^2}{102.39}$$

$$+ \frac{(98-102.39)^2}{102.39} + \frac{(42-40.77)^2}{40.77} + \frac{(8-6.84)^2}{6.84}$$

$$= 1.87$$

6. Do not reject H_0.

7. Conduct future analysis assuming that the weights are normally distributed.

This general test for goodness of fit can be used for any distribution. If we want to test the hypothesis that a collection of data is Poisson-distributed, for instance, we use the same procedure, gathering our expected proportions from the tables of the Poisson distribution. When you are checking for a Poisson distribution, the number of parameters estimated (w) from $k - w - 1$ df would be 1. We are estimating only the average number or μ.

AUTOPSY

A simple device can be used to shorten the number of calculations required in an analysis-of-frequency test. Notice that the testing statistic

$$\chi^2 = \sum \frac{(O_i - E_i)^2}{E_i}$$

is a group of individual items to be added together and compared to the region of rejection. The experienced analyst eyeballs the observed data as compared to the expected values and finds the largest individual differences. He then calculates that $(O_i - E_i)^2/E_i$ and compares it to the region of rejection. If it is not in the region of rejection, he looks for the next largest difference and calculates that $(O_i - E_i)^2/E_i$. Once he has calculated enough of these to make their sum fall in the region of rejection, he does not have to do any further calculations because the number can only get larger. To calculate the remaining elements is therefore redundant.

A common faux pas of the beginning analyst is to use something other than actual counts or frequencies in analysis-of-frequency tests. If the data observed are in the form of proportions or ratios,

TABLE 12-7.   Number of Respondents

| | Opinion of Product | | | |
Age	Favorable	Neutral	Unfavorable	Total
16–25	30	10	10	40
26–35	3	6	21	30
36–45	4	3	13	20
46 and over	3	1	6	10
	30	20	50	100

do not try to force them into a χ^2 analysis-of-frequency test. They just don't fit. One further word of caution on the subject: The *expected* number of observations must be at least 5 before you should attempt to use a χ^2 analysis of frequency. If any expected value is less than 5, categories should be combined to bring the expected value up to an acceptable level. To demonstrate this concept of having expected values of at least 5, consider the following problem.

A marketing manager has research data which indicate that a new product will not be acceptable. However, he thinks that the market for this product may be segmented by age. From a sample survey he has the data shown in Table 12-7.

This information could of course be translated into a contingency table test to determine whether the quality "opinion of product" is independent of the quality "age." We would hypothesize that the two qualities are independent and proceed with a χ^2 test:

$$\chi^2 = \sum \frac{(O - E)^2}{E} \quad \text{with } (r - 1) \cdot (c - 1) \text{ df}$$

The degrees of freedom here would be $(r - 1) \cdot (c - 1) = (4 - 1) \cdot (3 - 1) = 6$. Note, however, the expected number of observations for the cell "neutral opinion" and "46 and over." This number would be calculated as

$$\text{Expected number} = \frac{10}{100} \cdot 20 = 2$$

Similarly, the expected number for the cell "favorable opinion" and "46 and over" would be

$$\text{Expected number} = \frac{10}{100} \cdot 30 = 3$$

Both these expected numbers are less than 5. To work with this problem we could combine the age categories of "46 and over" and "36 to 45." The resulting category would be "36 and over," and we would combine the observations from the two original categories. Table 12-8 shows this combination.

TABLE 12-8. Number of Respondents

| | Opinion of Product | | | |
Age	Favorable	Neutral	Unfavorable	Total
16–25	20	10	10	40
26–35	3	6	21	30
36 and over	7	4	19	30
	30	20	50	100

All expected values in this table will be 5 or more, and we can proceed as originally planned. The degrees of freedom of our new contingency table would be $(r-1) \cdot (c-1) = (3-1) \cdot (3-1) = 4$ instead of the original 6 df.

SUMMARY

Analysis of Frequencies — Testing Statistic

$$\chi^2 = \sum \frac{(O_i - E_i)^2}{E_i}$$

1. Testing against presupposed proportions:

$$df = k - 1$$

2. Testing contingency tables (independence):

$$df = (r-1) \cdot (c-1)$$

3. Testing goodness of fit:

$$df = k - w - 1$$

where w is the number of estimated parameters.

PROBLEMS

1. A plant manager has three 8-hour shifts on his production line. He is questioning whether there are more accidents on some shifts than on others. If everything is under control, the three shifts should have the same frequency of lost-time accidents. He looks at the last 90 lost-time accidents and sees that 23 happened on shift 1, 35 on shift 2, and 32 on shift 3. Test the hypothesis that the frequency of accidents is the same on all three shifts. Use $\alpha = .05$.

2. A political pollster wants to determine whether there is a relationship between voters' income and their choice of candidate. He conducts a man-on-the-street interview campaign and comes up with the following cross-classification:

Candidate Preferred	Under $5,000	$5,000–$12,000	$12,001–$20,000	Over $20,000	Total
			Income of Voter		
Mr. Clean	13	15	17	10	55
Burly Bart	5	22	34	4	65
	18	37	51	14	120

Test the hypothesis of independence at $\alpha = .05$.

3. You are the head of the accounting department of a large trucking company. You are now working on a project to forecast tire expenses and need to know whether tire wear for your trucks is normally distributed. The following data are gathered from records:

Miles Used before Replacement	No. of Tires
18,000–19,000	25
19,000–20,000	76
20,000–21,000	120
21,000–22,000	58
22,000–23,000	34
	313

Calculating the mean $\bar{x}$ and the variance S^2, you find that

$$\bar{x} = \frac{\Sigma fx}{n}$$

$$= \frac{6,416,500}{313}$$

$$= 20,500$$

and

$$S = \sqrt{\frac{n\Sigma fx^2 - (\Sigma fx)^2}{n(n-1)}}$$

$$= \sqrt{\frac{313(131,908,250,000) - (6,416,500)^2}{97,656}}$$

$$= \sqrt{1,185,897}$$

$$= 1089$$

Test the hypothesis that the data are normally distributed with $\alpha = .05$.

4.   The district captain for a political party wants to know whether party affiliation is in any way connected with the age of the registered voter. A sample shows the following results:

Age	Republican	Democrat	Independent
30 or under	10	20	20
Over 30	60	50	40

Test the appropriate hypothesis at $\alpha = .05$.

5.   You are being investigated by the Civil Rights Commission, who say you are violating the standards of hiring minorities proportionate to the surrounding population. The area population proportions are 15 percent black, 20 percent Chicano, 5 percent American Indian, and 60 percent white. Your records show the following for your employees:

Whites	132
American Indians	8

Blacks	10
Chicanos	50

Conduct the appropriate test of hypothesis to defend yourself at an α level of 5 percent.

6. The only profit that Harry Hardrock makes from his defunct silver mine is selling ore samples to tourists at the front office. He sells these samples in small buckets of three different colors. Harry thinks there may be some relationship between the color of the pail and the sex of the tourist. He observes the following:

Color of	Sex of Tourist		
Bucket	Man	Woman	Can't Tell
Red	60	40	20
Black	15	25	10
Lavender	5	5	20

Test the appropriate hypothesis at $\alpha = .10$.

7. Complete the problem shown in Tables 12-7 and 12-8.

8. A random selection of time intervals over a number of days has shown the following number of incoming calls per two-minute period:

5	6	15
3	9	11
5	5	5
4	10	0
0	5	3
7	7	3
4	1	2
2	4	5
5	6	6
6	2	4

Test the hypothesis that the phone calls per two-minute period follow a Poisson distribution. (*Hint:* Find the average number of phone calls and refer to the Poisson tables.)

9.   A researcher is conducting experiments with three different treatments for a certain disease. He has gathered the following data from his experiments on 200 patients who were completely cured, were not changed, or got worse:

Patient Reaction	Treatment Received		
	A	B	C
Cured	40	21	24
No change	30	21	9
Worse	30	18	7

Test the appropriate hypothesis at a 5 percent level of significance.

10.   A police commissioner believes that 20 percent of the police calls are false alarms or unneeded, 70 percent are needed but noncritical, and 10 percent are critical (life or death situations). A sample of 80 calls shows that 20 were unnecessary, 20 were critical, and 40 were necessary but not critical. Was the commissioner's original assumption correct?

SELECTED REFERENCES

Dixon, W. J. and F. J. Massey, *Introduction to Statistical Analysis.* Third edition. New York: McGraw-Hill, 1969.

Guenther, William C. *Concepts of Statistical Inference.* Second edition. New York: McGraw-Hill, 1973. Especially Chapter 6.

Yamane, Taro. *Statistics: An Introductory Analysis.* Third edition. New York: Harper & Row, 1973. Especially Chapter 21.

CHAPTER 13
ASSOCIATION ANALYSIS

OBJECTIVES: After studying this chapter and working the problems, you should be able to:

Illustrate the concept of correlation *using a scatter diagram.*

Calculate the coefficient of correlation of data.

Interpret the coefficient of correlation of data.

Discuss the validity of a cause-and-effect assumption.

Compute and explain a regression equation.

Compute and explain the meaning of the standard error of the estimate.

Identify and explain the significance of the components of a time series.

SOMETIMES WE ARE concerned with the relationship or association between variables. This chapter deals with some of the techniques available for analyzing how variables are associated.

CORRELATION ANALYSIS

We tend to do informal correlation analyses every day. The term *correlation* has roughly the same meaning in statistical analysis as it does in general conversation. We hear statements like, "Boy, every time I speed there's a dozen policemen around." This is to say that the speed of your car is "correlated" with the number of policemen in the immediate vicinity. We also may commonly hear, "Every time I don't have time to study we get a pop quiz. If I do study, we never have a quiz." This is to say or imply that the number of hours spent studying is "correlated" with the number of quizzes that are given. And sometimes we hear a friend moan, "The harder I work, the behinder I get!" This implies that by some sort of black magic the amount of work one does is "correlated" with the amount of work left to be done.

These are informal correlation inferences that are commonly made. Obviously, when we make such statements they merely *seem* to be that way; if we watched carefully, we would notice that they don't generally hold true. We need, for our real-life situations, a procedure for measuring how well two items or characteristics are correlated. This procedure can be of value to us in a number of ways. For instance:

> If we could determine that the quality of our products is closely correlated with the amount of light in the assembly area, we might want to investigate the lighting in all our assembly areas.
>
> If we could determine that our sales volume is closely correlated with the amount of advertising done, we might want to look more closely at our advertising budget.
>
> If we could determine that there is a correlation between promotions and how closely an employee lets the boss beat him at golf, we might want to investigate other job opportunities.

Let's begin to formalize our concept of correlation. First of all, we begin our measure of correlation between two characteristics by gathering a sample of observations of the two characteristics. In the examples just given we would gather observations on the following:

> The number of defects that occurred when the lighting was at each of a number of levels of footcandles
>
> In each of a number of periods, the sales volume versus amount of advertising
>
> For each of a number of employees, the employee's salary versus how close the boss beats him at golf

In other words, we are gathering *pairs* of observations. We then compute from these pairs of observations a coefficient that we will denote by r. Before we look at the formula for computing r, let's examine some of its characteristics.

The correlation coefficient r will always be between -1 and $+1$. When the two characteristics of concern (we'll call them variable x and variable y) have a correlation coefficient that approaches -1, it means that whenever x increases, y almost always decreases and vice versa. When x and y have a correlation coefficient of close to $+1$, it means that they move predictably in the same direction or when x increases, y almost always increases and vice versa. When the correlation coefficient is close to zero, it means that no matter what direction x moves, y is liable to move in any direction and vice versa. In other words, the two variables are uncorrelated and one cannot be used to predict the other. This relationship is illustrated in the scatter diagrams of Figure 13-1. A scatter diagram shows the relative positions of *pairs* of observations; the horizontal axis represents the values of one characteristic and the vertical axis represents the values of the other. The formula for the correlation coefficient is

FIGURE 13-1.   Scatter Diagrams for Two Variables

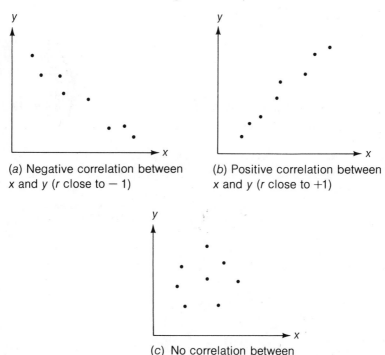

(a) Negative correlation between x and y (r close to -1)

(b) Positive correlation between x and y (r close to $+1$)

(c) No correlation between x and y (r close to 0)

SO&B 13-1: $$r = \frac{n\Sigma xy - \Sigma x\Sigma y}{\sqrt{[n\Sigma x^2 - (\Sigma x)^2][n\Sigma y^2 - (\Sigma y)^2]}}$$

Example 1: A college placement officer is beginning to wonder whether a student's grade point average has anything to do with the job offers the student receives. He decides to take a small sample of six graduating seniors and find the correlation coefficient between grade point average (GPA) and the number of job offers received. Table 13-1 shows the data from the sample (and the subsequent calculations he will need).

TABLE 13-1. Sample Data for GPA and Job Offers

Student	GPA (x)	No. of Offers (y)	x^2	y^2	xy
1	3.90	7	15.21	49	27.30
2	2.50	2	6.25	4	5.00
3	3.00	4	9.00	16	12.00
4	4.00	8	16.00	64	32.00
5	2.00	0	4.00	0	0
6	3.80	6	14.40	36	22.80
Sums (Σ)	19.20	27	64.90	169	99.10

Using SO&B 13-1 and the data from Table 13-1, he would then calculate

$$r = \frac{n\Sigma xy - \Sigma x\Sigma y}{\sqrt{[n\Sigma x^2 - (\Sigma x)^2][n\Sigma y^2 - (\Sigma y)^2]}}$$

$$= \frac{6(99.10) - 19.20(27)}{\sqrt{[6(64.90) - (19.20)^2][6(169) - (27)^2]}}$$

$$= \frac{76.20}{\sqrt{(20.76)(285)}}$$

$$= \frac{76.20}{\sqrt{5916.60}}$$

$$= \frac{76.20}{76.91}$$

$$= .99$$

This result should indicate to the analyst that there is a strong correlation between grade point average and the number of job offers received.

Example 2: Supposeth that our sad Danish prince is still pondering on why he has to suffer so many slings and arrows. He decides that perhaps it isn't outrageous fortune that causeth all his woes. He takes a sample of eight weeks and records for each week the number of showers (x) he took and the number of slings and arrows (y) he suffered. The resulting data (and the subsequent calculations he will need) are shown in Table 13-2.

TABLE 13-2.   Showers Taken and Slings and Arrows Suffered

Week	Showers (x)	S&A's (y)	x^2	y^2	xy
1	3	14	9	196	42
2	6	8	36	64	48
3	10	6	100	36	60
4	12	4	144	16	48
5	0	20	0	400	0
6	7	10	49	100	70
7	14	2	196	4	28
8	4	15	16	225	60
	56	79	550	1041	356

Using SO&B 13-1 and the data from Table 13-2, he then calculates

$$r = \frac{n\Sigma xy - \Sigma x \Sigma y}{\sqrt{[n\Sigma x^2 - (\Sigma x)^2][n\Sigma y^2 - (\Sigma y)^2]}}$$

$$= \frac{8(356) - 56(79)}{\sqrt{[8(550) - (56)^2][8(1041) - (79)^2]}}$$

$$= \frac{-1576}{\sqrt{(1264)(2087)}}$$

$$= \frac{-1576}{1624}$$

$$= -.97$$

This should tell the prince that there is a strong negative correlation between the number of showers he takes and slings and arrows he has to suffer. Perhaps it is not outrageous fortune that causes people to be nasty to him, but indeed something smelling rotten in the state of Denmark.

THE CAUSE-AND-EFFECT SYNDROME

There is a tremendous impulse for the beginning analyst to assume a cause-and-effect relationship between any two variables that show a strong correlation. This is a dangerous impulse that must be treated with a great deal of caution. For instance, there has been an analysis showing a strong positive correlation between teachers' salary increases and liquor consumption. This is not to say, though, that every time teachers get a raise they spend it on liquor! According to the *Wall Street Journal* (January 12, 1972), Edward R. Dewey has reported a correlation between the abundance of red squirrels and the level of steel production. But this does not imply that red squirrels cause steel production or vice versa.

When you find a strong correlation between two items, you must rely upon your knowledge and experience with these two items before making any inferences about cause and effect. Keep in mind that the calculation only tells you about the degree of correlation . . . it implies nothing about cause and effect. The same discussion of cause-and-effect relationships holds for regression analysis, which we will take up next.

REGRESSION ANALYSIS

With correlation analysis we were concerned with finding whether a relationship existed between two variables. Regression analysis is an attempt to find the relationship between variables to the extent that we can estimate the value of one by knowing the values of the others. We will limit ourselves here to what is called *simple, linear regression,* which is the analysis of two variables that have a *straight-line* relationship. In other words, they have a relationship of $y = a + bx$, where y and x are the two variables of concern and a and b are two constants that must be estimated.

The purpose of a regression equation is to predict or estimate one variable (called the *dependent variable*) from our knowledge of another variable (called the *independent variable*). A simple illustration would be the toll we pay at a certain toll bridge. Suppose the toll is $0.50 plus $0.10 per ton of the vehicle's weight. The variable to be estimated (the dependent variable) will be denoted by y. The variable used to estimate y (the independent variable) will be the weight in tons of the vehicle, denoted by x. We can "predict" the value of y, then, by putting the correct value of x into the formula

$$y = .50 + .10x$$

where x is the weight of the vehicle in tons.

The variables x and y have a linear relationship, and we can predict perfectly if we know the value of x. The equation

$$y = .50 + .10x$$

can be considered a regression equation. In this extreme value we know exactly the value of a and b in the general format

SO&B 13-2: $y' = a + bx$

Specifically, we know that $a = .50$ and $b = .10$. We will be attempting to find regression equations of the form of SO&B 13-2, but we will have to estimate a and b instead of knowing them exactly. Again the general purpose is to predict or estimate y (which we don't know) from the value of x (which we do know or can obtain).

As with correlation analysis, we gather data in pairs (a value for y and a value for x) from historical data or observations. To estimate a and b for a linear relationship we use *the method of least squares*, which guarantees that the resulting linear regression equation has less deviation from our actual observed data points than any other straight line we can devise.

To estimate b from these pairs of observations we calculate

SO&B 13-3:
$$b = \frac{n\Sigma xy - \Sigma x \Sigma y}{n\Sigma x^2 - (\Sigma x)^2}$$

We find our estimate of a by

SO&B 13-4:
$$a = \bar{y} - b\bar{x}$$

Example 3: We previously found the correlation coefficient r for the two variables "grade point average" and "number of job offers." Let's return to that situation. Suppose that the college placement officer is interested in predicting the number of job offers a student will receive. He can make a regression equation of the form $y' = a + bx$, where y' is the predicted number of job offers and x is the student's grade point average. Using SO&B 13-3 and SO&B 13-4 for the data contained in Table 13-1, we would calculate

$$b = \frac{n\Sigma xy - \Sigma x \Sigma y}{n\Sigma x^2 - (\Sigma x)^2}$$

$$= \frac{6(99.10) - 19.20(27)}{6(64.90) - (19.20)^2}$$

$$= \frac{76.20}{20.76}$$

$$= 3.67$$

$$a = \bar{y} - b\bar{x}$$

$$= \frac{27}{6} - 3.67 \left(\frac{19.20}{6} \right)$$

$$= -7.24$$

Our estimate of the regression equation, then, is

$$y' = -7.24 + 3.67x$$

where x is the student's grade point average. To estimate the number of job offers a student would receive, that student's grade point average would be put into the preceding equation. If a student had a grade point average of 3.25, then we would estimate the number of job offers he would receive as

$$y = -7.24 + 3.65x$$

$$= -7.24 + 3.65(3.25)$$

$$= 4.62$$

VALIDITY OF THE REGRESSION EQUATION

We cannot accept a regression equation carte blanche. The first limitation is that we cannot assume that the regression equation extends indefinitely beyond the range of the sample data used to construct the regression equation. In a previous example, the values of x in Table 13-1 can be seen to be 2.00 to 4.00. We cannot automatically assume that our regression equation is still valid for grade point averages below 2.00 or above 4.00. As a matter of fact, if we attempted to use our regression equation for estimating the job offers for a student with a 1.00 grade point average, we would find that

$$y = -7.24 + 3.65x$$

$$= -7.24 + 3.65(1.00)$$

$$= -3.59$$

And to say that a person is going to receive a negative number of job offers defies definition. The rule is that the regression equation tells us only about the range of data we used in the construction of that equation. If the analyst projects beyond those data, he should be aware that he is on his own and should carefully think through the validity of this extension before he proceeds.

The second limitation requires the realization that the calculation of estimates for a and b in $y' = a + bx$ is purely a mechanical process. What this means to you is that the estimation of $y' = a + bx$ for any two variables does not tell you whether or not the resulting regression equation is *usable*. I can select values for x and y from a table of random numbers and calculate from that a regression equation. But I seriously doubt whether I could make any *reliable* estimates of y from that regression equation. The point is that even though we can mechanically crank out a regression equation for any two variables, this does not necessarily help us — because our *purpose* in making a regression equation is to give us a way to estimate y accurately from our knowledge of some other variable, x.

If we are to be estimating y from x accurately, we must be able to determine whether our regression equation is valid. This validity is measured by two items. One of these is the correlation coefficient, which tells us how well the two variables are related. If two variables have a correlation coefficient close to $+1$ or -1, we would expect a regression equation for these two variables to be reliable or valid. Conversely, if the two variables x and y have a correlation coefficient close to zero, we would not expect a regression equation between the two to yield reliable estimates of y. (The correlation of the variables "number of job offers" and "grade point average" was found to be .99. This indicates that our regression equation is valid.)

The other item by which we can measure the validity of the regression equation is the *standard error of estimate*, $S_{y \cdot x}$. The standard error of estimate is a measure of how much our regression equation varies from the actual data points in our sample. Let's look back at the data in Table 13-1 and the resulting regression equation, $y' = -7.24 + 3.65x$. If we were to estimate, from our regression equation, the number of job offers expected for a GPA of 3.90, we would calculate

$$y' = -7.24 + 3.65x$$
$$= -7.24 + 3.65(3.90)$$
$$= 7.00$$

This result corresponds exactly with our sample data for the first student, who had a GPA of 3.90 and received seven job offers.

If we were to estimate, from our regression equation, the number of job offers expected for a GPA of 2.50, we would calculate

$$y' = -7.24 + 3.65x$$

$$= -7.24 + 3.65(2.50)$$

$$= 1.89$$

This result varies slightly from the observation for the second student, who had a GPA of 2.50 and received two job offers. If we continued this procedure, we could generate the data displayed in Table 13-3.

TABLE 13-3. Calculation of Standard Error of Estimate ($S_{y \cdot x}$) for GPA Versus Number of Job Offers

Student	GPA (x)	Number of Offers Observed (y)	Computed (y')	(y − y')	(y − y')²
1	3.90	7	7.00	0	0
2	2.50	2	1.89	.11	.0121
3	3.00	4	3.71	.29	.0841
4	4.00	8	7.36	.64	.4096
5	2.00	0	0.06	−.06	.0036
6	3.80	6	6.63	−.63	.3969
					.9063

From the data in Table 13-3 we can calculate $S_{y \cdot x}$, the standard error of estimate, using SO&B 13-5:

SO&B 13-5:
$$S_{y \cdot x} = \sqrt{\frac{\Sigma(y - y')^2}{n - 2}}$$

For our current example,

$$S_{y \cdot x} = \sqrt{\frac{.9063}{4}}$$

$$= \sqrt{.23}$$

$$= .48$$

We use and interpret the standard error of estimate in the same way we used and interpreted the standard deviation in Chapter 4.

From our results here we would say that with $r = .99$ and $S_{y \cdot x}$ as small as .48, our regression equation of $y' = -7.24 + 3.65x$ is a reliable estimator of the number of job offers for GPAs between 2.00 and 4.00.

The formula for $S_{y \cdot x}$ given in SO&B 13-5 is a definitional formula. It is generally much easier to use the computational form

SO&B 13-6: $$S_{y \cdot x} = \sqrt{\frac{\Sigma y^2 - a \Sigma y - b \Sigma xy}{n - 2}}$$

which utilizes the data that have already been calculated for the correlation coefficient and the regression equation.

> **Example 4:** Supposeth our Danish prince is now interested in predicting the number of slings and arrows he will have to suffer during the week. From the data in Table 13-2, and using SO&B 13-3 and SO&B 13-4, we can construct a regression line for him as follows:
>
> $$b = \frac{n\Sigma xy - \Sigma x \Sigma y}{n\Sigma x^2 - (\Sigma x)^2}$$
>
> $$= \frac{8(356) - 56(79)}{8(550) - (56)^2}$$
>
> $$= \frac{-1576}{1264}$$
>
> $$= -1.25$$
>
> $$a = \bar{y} - b\bar{x}$$
>
> $$= \frac{79}{8} - (-1.25)\left(\frac{56}{8}\right)$$
>
> $$= 9.875 + 8.750$$
>
> $$= 18.63$$

The resulting regression equation would be $y' = 18.63 - 1.25x$, where x is the number of showers he takes per week.

To determine the reliability of the regression equation we have already found that $r = -.97$ and, using SO&B 13-6, we find the $S_{y \cdot x}$ to be

$$S_{y \cdot x} = \sqrt{\frac{1041 - 18.63(79) + 1.25(356)}{6}}$$

$$= \sqrt{\frac{14.23}{6}}$$

$$= 1.54$$

The correlation coefficient and the standard error of estimate indicate that we can reliably predict the number of slings and arrows our prince has to suffer based on the number of showers he takes.

TIME SERIES

Sometimes we are concerned about the association of a variable with time. Any comparison of a variable over some range of time is called a *time series*. The time series is so much a part of the business world that almost any cartoon depicting a business situation will show in the background a chart of sales compared to time.

Time series are used extensively in forecasting, where the purpose is to identify how the variable (sales, for instance) behaves over a period of time and then use that identified pattern to project into the future. With time-series analysis we attempt to break that "identified pattern" into four parts, find how each part behaves, and then project into the future by putting the four parts back together. The four parts we attempt to distinguish are these:

1. Long-term trend — the general tendency to move in a certain direction over a long period of time; we often estimate this trend with a straight line.

2. Seasonal variations — the recurring ups and downs within a one-year period, such as Christmas sales.

3. Cyclical fluctuations — that part of the time-series movement which is due to overall business or economic cycles.

4. Irregular movements — the innumerable factors that can disrupt normal operations, such as strikes, floods, earthquakes, riots, accidents. These unpredictable disruptions that affect a time series are classified as "irregulars."

FIGURE 13-2.   Sales for Captain Candy Cereal: 1969–1974

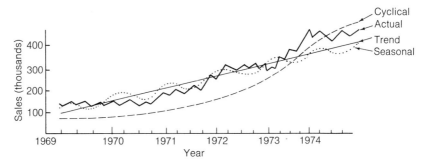

The analyst is primarily concerned with the long-term trend if he is forecasting on an annual basis; this trend is often estimated from historical data by a regression line, using the years for the *x*'s and the variable of concern (sales, for instance) as the *y*'s. If the analyst is concerned about an estimate within a year, then he must investigate the effect of seasonal variations. Cyclical fluctuations are generally predicted by an economist. The irregulars are unpredictable, of course, by nature.

A time series is shown graphically in Figure 13-2. The projection of sales into 1975 and beyond would be accomplished by extending each of the three components and then combining their effects. The fourth component, irregulars, has to be handled as it arises.

SUMMARY

The would-be analyst must be extremely cautious in making cause-and-effect assumptions based on association analysis.

Correlation Coefficient

SO&B 13-1: $r = \dfrac{n\Sigma xy - \Sigma x \Sigma y}{\sqrt{[n\Sigma x^2 - (\Sigma x)^2][n\Sigma y^2 - (\Sigma y)^2]}}$ $-1 \leqslant r \leqslant +1$

Regression Analysis

SO&B 13-2: $y' = a + bx$

SO&B 13-3: $b = \dfrac{n\Sigma xy - \Sigma x \Sigma y}{n\Sigma x^2 - (\Sigma x)^2}$

SO&B 13-4: $a = \bar{y} - b\bar{x}$

Time Series Components

1. Long-term trend

2. Seasonal variation

3. Cyclical fluctuation

4. Irregular movement

PROBLEMS

1. Describe two situations in which correlation and regression analysis would be useful. Explicitly define x and y (the independent and dependent variables).

2. A production manager is concerned about the possible association between machine speed on a certain machine and the number of defects turned out by that machine. He gathers the following random sample of machine speeds and defects:

Observation	Machine Speed (rpm)	No. of Defects
1	450	15
2	425	14
3	350	12
4	300	9
5	375	10
6	400	13

a. Make a scatter diagram of these data.

b. Calculate the correlation coefficient.

c. Calculate the regression equation.

d. Calculate the standard error of estimate.

e. How reliable do you think the regression equation from part (*c*) is? Use your answers from parts (*b*) and (*d*) to make this determination.

f. Estimate the number of defects that would be expected if the machine speed were set at 410 rpm.

3.   A union steward suspects that there is a relationship between the amount of lighting and industrial accidents. He gathers the following data as he varies the lighting on the assembly line:

Observation	Lighting (in footcandles)	Accidents Per Week
1	1750	1
2	1600	0
3	1500	2
4	1000	6
5	1300	4
6	1200	5
7	1400	3

   a.   Make a scatter diagram of these data.

   b.   Calculate the correlation coefficient.

   c.   Calculate the regression equation.

   d.   Calculate the standard error of estimate.

   e.   How reliable do you think the equation in part (*c*) is?

4.   Little Red Riding Hood believes that the wolf may be attacking her because he wants the goodies in her picnic basket. She experiments with this idea by varying the amount of goodies in her basket. She gathers the following data:

Week	Weight of Basket (lb)	No. of Attacks
1	2	3
2	3	6
3	9	2
4	1	0
5	8	5
6	10	5
7	2	6
8	3	1
9	5	4
10	8	0

a. Calculate the correlation coefficient.

b. Calculate the regression equation.

c. Calculate the standard error of estimate.

d. How reliable do you think your answer to part (*b*) is?

5. A company has the following record of sales by quarter for 1970 through 1974:

Quarter	1970	1971	1972	1973	1974
1st	1000	1100	1250	1400	1425
2nd	1500	1550	1700	1800	1875
3rd	1000	1000	1300	1375	1400
4th	2000	2150	2250	2275	2400
	5500	5800	6500	6850	7100

a. Plot the quarterly data on a chart so that you can visually inspect this time series. Using this chart, estimate the company's sales for 1976.

b. Find the trend of this time series by calculating the regression equation of total year's sales versus year. (*Hint:* Code the years 1970 = 0, 1971 = 1, 1972 = 2, and so on for easier calculation.)

c. Estimate the company's sales for 1976 using your answer from part (*a*).

d. Compare this answer to that for part (*a*).

e. Do this company's sales seem to have any seasonal pattern? Why or why not?

6. A hypothetical regional traffic department has records that show the number of fatal traffic accidents per year. They want to find whether there is any relationship between fatal accidents and (1) the amount of required safety equipment on cars measured in dollars and (2) the maximum freeway speed limit. They have the following data for five years:

Fatal Accidents Per 100,000 Population	Value of Safety Equipment ($)	Maximum Freeway Speed
34.1	76	70
37.2	105	80
35.9	105	75
30.6	110	60
29.1	250	55

a. Find the correlation coefficient between accidents and safety equipment.

b. Find the correlation coefficient between accidents and maximum freeway speed.

c. Should the answers to parts (*a*) and (*b*) give any useful information to the regional traffic department? If so, what?

d. Calculate a regression equation for estimating the traffic fatalities (*y*) from the maximum freeway speed (*x*).

e. Could you reasonably assume a cause-and-effect relationship between fatal accidents and maximum freeway speed? Why or why not?

SELECTED REFERENCES

Bryant, Edward C. *Statistical Analysis.* Revised edition. New York: McGraw-Hill, 1966.

Guenther, William C. *Concepts of Statistical Inference.* Second edition. New York: McGraw-Hill, 1973. Especially Chapter 8.

Miller, Irwin and John E. Freund. *Probability and Statistics for Engineers.* Englewood Cliffs, N.J.: Prentice-Hall, 1965. Especially Chapter 12.

Spurr, William A. and Charles P. Bonini. *Statistical Analysis for Business Decisions.* Revised edition. Homewood, Ill.: Irwin, 1973. Especially Chapters 16, 19, 20.

Tanur, Judith M., Frederick Mosteller, William H. Kruskal, Richard F. Link, Richard S. Pieters, and Gerald R. Rising. *Statistics: A Guide to the Unknown.* San Francisco: Holden-Day, 1972.

Yamane, Taro. *Statistics: An Introductory Analysis.* Third edition. New York: Harper & Row, 1973. Especially Chapters 12, 13, 14, 15.

CHAPTER 14
INDEX NUMBERS

OBJECTIVES: After studying this chapter and working the problems you should be able to:

Investigate and report on the source, method of calculation, and meaning of a commonly used index number.

Calculate a simple index, a weighted aggregative price index using the Laspeyres method, and a weighted aggregative price index using the Paasche method.

INDEX NUMBERS ARE special ratios whose purpose usually is to show *relative* changes over time. We are commonly exposed to such index numbers as the cost-of-living index (formally, the consumer price index) and the wholesale price index. These index numbers are an attempt to measure changes in prices over time. Before we look at ways to calculate index numbers, we need to know some basic terminology: The *base period* is that time period from which the changes will be measured; it may be one year or a number of years averaged together. The following notation is used in discussions of index numbers:

p_0 = the price of a commodity in the base year

p_n = the price of a commodity in any given year (n) other than the base year

q_0 = the quantity of a commodity sold or purchased in the base year

q_n = the quantity of a commodity sold or purchased in any given year (n) other than the base year

I_n = the value of the index in any given year (n)

We will be using this terminology and notation in the discussion of index numbers. In their simplest form, index numbers contain only one variable; these are called *simple* index numbers. If more than one variable is involved, it is called a *composite* index number. We will look first at simple index numbers.

SIMPLE INDEX NUMBERS

Index numbers, in general, are used because the raw data are too complex for us to analyze and detect the amount of change from

TABLE 14-1. Production of Mineral A:
1968–1975 (Base Year 1970)

Year	Production (tons)	Index (I_n)
1968	250,000	83
1969	275,000	92
1970	300,000	100
1971	285,000	95
1972	270,000	90
1973	290,000	97
1974	310,000	103
1975	315,000	105

time period to time period. To gain insight into relative changes in an item we could create a simple index for it. To calculate the simple index number for quantities we merely form the ratio of the amount of the variable in the given year to the amount of the variable in the base year. This ratio is usually multiplied by 100 to give a base of 100. Hence if we are measuring changes in quantity of a certain item, the index for any given year is

SO&B 14-1: $$I_n = \frac{q_n}{q_0} \cdot 100$$

Table 14-1 shows the production in tons of a hypothetical mineral and the resulting simple index, using 1970 as a base period. An index is translated as a percentage of the base year. For instance, in Table 14-1 the index value for 1975 is translated as "the production in 1975 was 105 percent of the production in 1970."

COMPOSITE INDEX NUMBERS

When we are measuring relative changes in complex situations, we usually combine items into one index number. In making a composite index we weight the relative importance of each item that goes into the index. If we are measuring relative changes in price, we can weight the price of each item by the quantity that was used. If we are measuring relative changes in quantity, we can weight the quantity of each item by the price of that item.

We can weight a composite index number in one of two ways: the *Laspeyres method* uses the base year weights; the *Paasche method* uses the given year weights. The following formulas are used to calculate a *weighted aggregative price index* (see the beginning of this chapter for a a review of the symbols used):

SO&B 14-2: $I_n = \dfrac{\Sigma p_n q_0}{\Sigma p_0 q_0} \cdot 100$ (Laspeyres method)

SO&B 14-3: $I_n = \dfrac{\Sigma p_n q_n}{\Sigma p_0 q_n} \cdot 100$ (Paasche method)

Notice that in SO&B 14-2 we use the quantities in the base year, q_0, as weights while in SO&B 14-3 we use the quantities in the given year, q_n, as weights. To demonstrate the calculation of composite

TABLE 14-2. Price and Quantity of
Ingredients Used: 1974–1975

Ingredient	Price		Quantity	
	1974	1975	1974	1975
Pork (lb)	$0.75	$0.70	2000	3000
Garlic (oz)	0.05	0.10	1000	1200
Sawdust (bushels)	3.00	3.00	1	2

index numbers using SO&B 14-2 and SO&B 14-3, consider the following problem.

A sausage manufacturer uses three ingredients in his product: pork, garlic, and sawdust. He is interested in watching the relative movements of the prices he pays for these ingredients. He has the data shown in Table 14-2.

If the manufacturer chose 1974 as the base year, he would put the data in terms of p_n's and q_n's as shown in Table 14-3.

TABLE 14-3. Calculation of Price Index for
Sausage Ingredients (Base Year 1974)

Item	p_0	q_0	$p_0 q_0$	p_1	q_1	$p_1 q_0$	$p_1 q_1$	$p_0 q_1$
Pork	.75	2000	1500	.70	4000	1400	2800	3000
Garlic	.05	1000	50	.10	800	100	80	40
Sawdust	3.00	1	3	3.00	2	3	6	6
Sums (Σ)			1553			1503	2886	3046

If he were to use the Laspeyres method (SO&B 14-2), he would calculate the index for 1975 as

$$I_1 = \frac{\Sigma p_n q_0}{\Sigma p_0 q_0} \cdot 100$$

$$= \frac{1503}{1553} \cdot 100$$

$$= 97$$

This result would be translated as: "The quantities purchased in 1974 (the base year) would cost 3 percent less if purchased in 1975." If the manufacturer were to use the Paasche method (SO&B 14-3), he would calculate the index for 1975 as

$$I_1 = \frac{\Sigma p_n q_n}{\Sigma p_0 q_n} \cdot 100$$

$$= \frac{2886}{3046} \cdot 100$$

$$= 95$$

This result would be translated as: "The quantities purchased in 1975 would cost 5 percent less in 1975 than they would have in 1974."

In both cases the index for the base year (1974) would have been 100; the index for a single base year is always 100. Note that the unit of measure does not have to be the same for all the commodities. A weighted aggregative index can accommodate such diverse measures as price per ounce and price per bushel.

You should pay careful attention to the difference in the translation of the two methods (Laspeyres versus Paasche). As in the example, the two methods will give different numerical answers when the quantities have changed in their relationship between the base year and the given year. The more drastic the change, the more difference there will be between the two methods. If the quantities have remained relatively the same, the two methods will yield identical results.

The Laspeyres method (using base year as weights) is used more often than the Paasche method (using given year as weights). One reason for this preference is that the Laspeyres method retains a constant base or denominator that makes the calculations easier for additional years. However, discussions about which method is better are purely academic because each method's value depends entirely on what the analyst is trying to measure. If he is primarily concerned about price movements for the base-year quantities, the Laspeyres method should be used. If he is primarily concerned about price movements involving the quantities being used now, the Paasche method should be used.

If you are presented with an index number rather than calculating it, you need to know how (that is, by what method) the index number was calculated. How you can use that index number depends on its method of calculation.

USES OF INDEX NUMBERS

As mentioned, index numbers are used primarily to measure relative change. For example, index numbers are used to measure the economic strength of the nation and to predict future sales. The consumer price index is used in labor relations for wage negotiations, but it is also often used to *deflate* prices. Deflating prices means reducing prices or income to constant dollars relative to some base year. In its simplest form, this amounts to dividing prices or income in a given year by the consumer price index for that year.

AN EXAMPLE: THE BUSINESS WEEK INDEX

One special index number used by many businesses is the Business
Week Index. The following article* was published to explain that
index. In reading the article watch for the concepts and terminology
we have used thus far.

THE INDEX SHOWS A STRONG RECOVERY

In the past month, the Business Week Index rose 1.3 percent and
now has added nearly 6½ points since the beginning of 1972, a sure
sign that the U.S. economy is making a strong, healthy recovery. Ever
since it was formulated for the issue of May 7, 1930 — a few months
after the magazine was born — the index has been one of the most
sensitive barometers of the well-being of the economy. It also has
been an accurate herald of the industrial production index issued
monthly by the Federal Reserve Board, itself a well-recognized indicator of business activity.

The rise in activity shown by the index — a statistical representation
— is a reflection of the real gains posted in wide areas of business. For
instance, since the beginning of the year, steel production has risen
30 percent, output of electricity has advanced 13 percent and rail
freight is up 6 percent. All of these are components of the index.

Now, the Business Week Index has been extensively revised to reflect what has happened to the structure of the economy since the
last major revision of the series in 1963. The base period has been
shifted from 1957–'59 to 1967, and the components of the index have
been expanded and reweighted.

For instance, the old index lumped truck production with auto
output. But the pattern of truck production differs markedly from
that of autos. There is more competition among materials, so this
has reduced the weight given to steel in the index. Coal did not seem
so important when the index was first revised in 1938, and it was
dropped out of the list. Now it is back. There have been other significant changes.

* Reproduced from *Business Week*, October 21, 1972, by special permission.
Copyright 1972 by McGraw-Hill, Inc.

An Early Warning

The Business Week Index is not intended to be a forecasting device, nor does it presume to provide a picture of the entire economy. Instead, it takes a candid snapshot of the economy's most dynamic area: industrial production. And, by limiting its focus, it gains the sensitivity to foreshadow the short-term swings and major turning points in the industrial sector that determine the direction and force of the economy as a whole. Because it picks up such trends so quickly, it has become part of an early warning system that businessmen and economists find uniquely valuable.

Pierre Rinfret, president of economic consultants Rinfret-Boston Associates, maintains a "tracking and warning system" for his clients, and he says: "On a weekly basis, the Business Week Index is the most valuable single barometer of the industrial sector of the American economy." Paul Markowski, chief economist for the Wall Street house of Weis, Voisin & Co., agrees. He previously was the index-builder for the Cleveland National Bank, and he now uses the Business Week Index—as do most other economic-statistics watchers—to "confirm my own sense of what's happening." He notes that "if the index is headed in a different direction from my appraisal, I review my appraisal, since the BW index has been a sensitive indicator of prospective production changes." The index also is submitted to 95 newspapers in 38 states and seven foreign countries.

The fact that the index is current makes it especially useful. The substantial collection of monthly data issued by the federal statistical agencies defines the underlying trends in business. But there can be a lag of weeks or more than a month before this information becomes available. The changes in the BW weekly index add up to the monthly change in industrial production, as measured by the Federal Reserve Board.

A New Lineup of 13

The components of the Business Week Index were specifically selected to represent the sum total of the industrial sector. In this present revision, the Economics Dept. of McGraw-Hill Publications and *Business Week* statisticians, after much experimentation with weekly economic data, have compiled a new, broader index of 13 components. In the revision process, two of the old components were displaced and five new ones were added. After elaborate computer analysis, a new set of weekly seasonal factors for each of the industries was calculated. And statistical weights assigned to each element of the index have been readjusted to reflect the changed roles of the industries covered.

All of the industrial sector is represented in the index – by direct measure, as in raw steel poured, or by broad indicators such as electricity output, trucking and rail carloadings, which ebb and flow with the pace of factory output. The production series includes machinery, oil refining, steel, cars, trucks, paper, paperboard, coal, lumber, and defense and space equipment. Together they comprise 46 percent of the weight of the index. Trucking and rail shipments account for 40 percent, with electricity making up the balance of 14 percent.

The weekly production series incorporated into the Business Week Index represent almost one-fourth of the total weight in the Federal Reserve's industrial production index. The other weekly series – rail freight, intercity trucking, and electricity – cover activity in industries, such as chemicals and textiles, that are not represented by an available weekly data. The index is also leavened with weekly extrapolations based on two monthly series: machinery because of its cyclical sensitivity, and defense and space because of its special role in the economy.

Old Favorites Change

The new composition of the index represents a considerable change from that of the index assembled a decade ago. For example, lesser statistical weights are now assigned to electric power output, steel, oil refinery runs, and paperboard:

> Electricity is now assigned a weight of 14 percent, in contrast with 29 percent in the 1963 series. One reason for the change is that air conditioning now plays such an important part in power consumption that weekly changes caused by the weather unduly influenced the index. There has also been a massive growth in the use of consumer appliances. The index is intended to react to changes in industrial activity, not to purely random or nonindustrial factors. Industrial and commercial consumption accounts for 67

percent of total power use, and thus should closely reflect productive activity. The index figure includes firm figures for 95 percent of investor-owned, municipal, and federal utility output; the other 5 percent in the figure is estimated. Steel continues to be the nation's single most important material, despite inroads by aluminum, plastics, and cement. Last year, steel shipments totaled $27-billion worth, and the industry employed 581,000 workers. However, tonnage of imported steel—outside the U.S. production figure—last year was three times that of the early 1960s. The new statistical weight of 4 percent reflects steel's changed position in the economic spectrum.

Nonetheless, steel remains a key factor. Any change in its output and market reverberates quickly through the entire economy. For instance, a steel strike begins to cut the output of a wide spectrum of industries within two months, even when inventories have been built up as a strike hedge. Steel is no longer measured by the ingot, which can vary from 5 to 30 tons, because not all steel comes in ingots. The material is now categorized as "raw steel."

Crude oil refinery runs are a count of the barrels of products—gasoline, fuel oil, lubricants, petrochemicals, etc.— that are refined from crude oil by all U.S. producers. In the petroleum business, a barrel is the unit of measurement for output, even though almost none of it ends up in that kind of container. A barrel is 42 gal., unlike the $32\frac{1}{2}$-gal. barrel of the liquor industry.

Paperboard is virtually a synonym for packaging. It is the packaging industry's most widely used intermediate material. Made from unbleached and sulfate pulp and secondary fibers such as waste paper, it winds up as containerboard—mostly boxes and filler material—or boxboard, the material for folding cartons, milk containers, and other types of food containers. Some form of paperboard, ranging from shipping crates and folding cartons to fancy packages, encases nearly every purchase by business or the public. Hence, paperboard production reacts quickly to shifts in market demand.

A Shuffle in Transportation

The segments of the transportation equipment field have been shuffled in the new Business Week Index to reflect changes in relative importance and also differences in production patterns:

> Auto production continues to be one of the dominant, as well as most sensitive, indicators of business activity. Production this year at an annual rate of more than 9-million cars chews up a long list of commodities, ranging from aluminum to zinc, and directly or indirectly affects millions of workers. The index figure is derived from the output of 44 U.S. assembly plants. Foreign cars, even those manufactured by partly owned U.S. subsidiaries, are not represented.

> Truck production is now calculated and listed separately in the index because its week-to-week pattern differs from that of autos and also because it is now a big enough factor to make a dent in the index. Last year, truck output reached 2.4-million units, almost double the output of a decade ago. Transportation equipment other than trucks has been eliminated from the index as a separate listing, but most of it now goes into the new category of defense and space equipment.

The index revision has eliminated other statistical lines, too, while adding some new ones:

> Construction contract awards for residential and other purposes have been dropped. It turned out that awards, available only monthly, show too wide a variation from month to month, either above or below the underlying trend of construction activity.

> Lumber has been added in place of construction contract awards, as a more direct production measure. Lumber accounts for 40 percent of the basic material in the construction of a single-family home, and nearly 15 percent in apartment-house construction. It also reflects potential production in the furniture industry and industrial packaging. Weekly data are collected from 225 mills that account for well over half of total U.S. production.

Coal production, a component eliminated from the index when the industry went into a decline, has been reinstated because of its revived importance in the economy. Coal's growing contribution to total energy consumption now accounts for 20 percent of the aggregate. Data are included for all mines that produce at least 1,000 tons a year. Last year, 548-million tons of bituminous were mined, 26 percent more than in 1961. Anthracite, however, was down to less than 9-million tons — about half the amount mined in 1961.

Paper production has been added as a new weekly series. This category includes paper used for newsprint, packaging, wrapping paper, shipping sacks, bags, tissues, commercial printing, book and magazine publishing, stationery, record forms, and computer printouts. These uses of paper respond readily to changes in general business conditions. Companies that report output represent 75 percent of the industry.

Distribution plays a greater role than ever before in determining the movement of the index. Trains and trucks are the main carriers of goods from factories, and data are available on a weekly basis. So these two categories have been given added weight in the revised index. Between them, they account for almost two-thirds of all intercity freight traffic, with pipelines and ships moving most of the remaining one-third.

Rail freight traffic is now incorporated in the index with a weight of 22 percent, in place of railroad carloadings, formerly weighted at 12 percent. Carloadings no longer accurately reflect the volume of freight; they have been declining because of a trend toward heavier loadings and longer trips. For instance, carloadings last year dropped 7 percent, but revenue ton-miles fell only 3.3 percent. In the past decade, the average distance per ton of freight hauled has risen 12 percent (to 505 mi.), and freight-car capacity has risen 23 percent to an average of 68 tons. Thus, revenue ton-miles on Class 1 railroads have been substituted as a more realistic measure of rail freight movement.

Intercity truck tonnage, also representing distribution, is now included in the index at a heavy weight of 18 percent, up from its former 2 percent. Intercity trucking covers a broad range of industries; trucks carried an estimated 422-billion freight ton-miles in 1971 and accounted for 22 percent of intercity freight moved throughout the country.

The Seasonal Patterns

Each of the individual series has a different seasonal pattern. Electricity peaks twice a year: in midwinter when short days lead to heavy usage of lighting, in midsummer when hot weather brings a jump because of air conditioning. Steel peaks in late spring and dips during the summer vacation period. Auto production rises in the spring because of seasonal improvement in sales, and again in the fall when new models come out. It declines sharply in late summer when the model changeover period arrives. All of these changes, and identifiable patterns in the other series, occur each year; the usual pattern can be calculated and the data adjusted to determine what the underlying rate of activity is, in absence of what are the purely seasonal influences.

After seasonal adjustments, the 13 components are combined into one index construction equation: $I = \Sigma Wc \, [qc^1/qc^\circ]$ where I equals the Business Week Index, Wc is the weight of the components, qc^1 equals the quantity of each component in the current period, and qc° equals the quantity of each component in the base period, 1967–100. What this equation does is to add together apples and oranges. For example, tons of raw steel, barrels of oil, and number of cars and trucks are blended into a single unit, the index number, by multiplying the proportionate weekly change in each component by its respective weight in the index. The index thus represents the sum total of all of the changes of all of the series, weighted by the size of the change in each series and its importance in the total. The index is calculated by a *Business Week* statistician on Tuesday of each week, incorporating the data for all 11 weekly series for the period ending 10 days previously. For instance, in the *Business Week* of Oct. 21, the index was computed on Oct. 17, for all series for the week ended Oct. 7.

The ability of the Business Week Index to pinpoint changes in production can be seen from its record of the past year or so. In July and August, 1971, industrial production softened, wiping out all the slight gain in factory output recorded in the sluggish business advance of the first half of the year. The July production figures became available in the middle of August; the August production numbers, in the middle of September. However, the Business Week Index began to slip early in July and continued to decline into early August, indicating that both months would register production declines. With the advent of the New Economic Policy in August, the level of business activity began to quicken. Late in August and in September, the index headed upward, again anticipating the change of direction in factory production. Over the longer pull, on a monthly basis, the Business Week Index coincides closely with the Federal Reserve's index of industrial production, seldom departing more than a point, above or below.

Before deciding upon the 13 components now in the index, *Business Week* statisticians reviewed many industries and many economic series to determine if they could be incorporated into the index. Most of the rejected series, however, suffered from shortcomings that precluded their use. For instance, radio-TV-phonograph was considered, but was rejected because there was no effective way to segregate foreign production—no small portion of the total—from the data on a weekly basis. Other industries that were explored but yielded no weekly figures included metals, textiles, air freight, and home appliances.

The monthly series, machinery and defense and space, were included despite the lack of weekly information, in order to cover an important segment of industry, at least on a monthly basis. Machinery accounts for over 17 percent of total industrial production as measured by the Federal Reserve; it is an important factor in the short-term fluctuations in the economy. The cyclical movements in machinery are keyed to the general business cycle, both for industrial (largely nonelectrical) and consumer (mostly electrical machinery) industries.

Defense and space equipment production includes all types of military hardware and aerospace equipment, some communications equipment, military aircraft, ships, and ordnance. Political considerations and budget requirements affect these industries, frequently in specific geographic areas—such as the 1970 cutback in aerospace, so heavily felt in California.

AUTOPSY

The would-be analyst should be aware that pitfalls exist in the use of index numbers. To avoid these he should:

1. Carefully consider and define the basic purpose of the index number.

2. Analyze the items to go into the index and select them according to importance and pertinence.

3. Select a base period that could be considered typical, avoiding years with long strikes, earthquakes, and the like.

SUMMARY

Index numbers are used to measure relative movements. A simple index is made up of only one item; a composite index is made up of more than one item.

SO&B 14-1: $\quad I_n = \dfrac{q_n}{q_0}$ (a quantity index for one variable)

SO&B 14-2: $\quad I_n = \dfrac{\Sigma p_n q_0}{\Sigma p_0 q_0} \cdot 100$ (the Laspeyres: a weighted aggregative price index with base year as weights)

SO&B 14-3: $\quad I_n = \dfrac{\Sigma p_n q_n}{\Sigma p_0 q_n} \cdot 100$ (the Paasche: a weighted aggregative price index with given year as weights)

PROBLEMS

1. The turnip production in Lower Slobovia was as follows:

Year	Production (tons)
1966	200
1967	210
1968	225
1969	140
1970	215
1971	230
1972	234
1973	234
1974	230
1975	240

      a. Create a simple quantity index for each year, using 1966 as a base period.

      b. Create a simple quantity index for each year, using the years 1966 to 1968 as a base period (find the average of the three years).

2.   What is the level of the current consumer price index?

3.   Find the source and an explanation of a commonly used index number (for example, the wholesale price index) and write a report on how it is derived and used. (The *BLS Handbook of Methods* describes index numbers calculated by the U.S. Department of Labor.)

4.   A manufacturer of flashlights uses bulbs, wire, plastic, and metal in his product. The prices and quantities used of these four items are shown here for three years:

Item	Price			Quantity		
	1973	1974	1975	1973	1974	1975
Bulbs (dozens)	$0.14	$0.15	$0.16	400	450	500
Wire (ft)	0.03	0.03	0.05	1000	1500	1200
Plastic (lb)	0.60	0.55	0.50	600	600	700
Metal (lb)	1.14	1.20	1.25	300	275	250

      a. Find the Laspeyres weighted aggregative price index for each of the years, using 1973 as the base period.

      b. Find the Paasche weighted aggregative price index for each of the years, using 1973 as the base period.

5.   Little Red Riding Hood has been keeping records on the quantity and prices of the items in the little basket she carries to Grandma's house. These prices and quantities are as follows:

Item	January		February		March		April	
	Price	Quantity	Price	Quantity	Price	Quantity	Price	Quantity
Cookies (dozens)	$0.75	3	$0.75	3	$0.85	4	$0.85	2
Sandwiches (each)	0.25	28	0.27	35	0.30	40	0.31	40
Soda pop (pints)	0.10	3	0.10	10	0.15	12	0.15	10

a. Find the weighted aggregative price index for her basket, using the Laspeyres method and March as a base period.

b. Find the weighted aggregative price index for her basket, using the Paasche method and March as a base period.

6. A candy bar contains chocolate, peanuts, and caramel. The prices and quantities of these ingredients are as follows:

		Price			Quantity	
Ingredient	1965	1970	1975	1965	1970	1975
Chocolate (pounds)	$0.08	$0.10	$0.12	100	150	200
Peanuts (bushels)	1.00	1.00	1.10	40	40	30
Caramel (pounds)	0.15	0.20	0.25	60	70	80

a. Find the weighted aggregative price index for each of the recorded years, using the Laspeyres method and 1965 as a base year.

b. Find the weighted aggregative price index for each of the recorded years, using the Paasche method and 1965 as a base year.

SELECTED REFERENCES

Spurr, William A. and Charles P. Bonini. *Statistical Analysis for Business Decisions.* Revised edition. Homewood, Ill.: Irwin, 1973. Especially Chapter 18.

Tanur, Judith M., Frederick Mosteller, William H. Kruskal, Richard F. Link, Richard S. Pieters, and Gerald R. Rising. *Statistics: A Guide to the Unknown.* San Francisco: Holden-Day, 1972. Especially pages 266–275.

U.S. Bureau of Labor Statistics. *BLS Handbook of Methods.* Washington, D.C.: U.S. Government Printing Office, 1971.

APPENDIX:
STATISTICAL
TABLES

TABLE A-1. Binomial Distribution (Cumulative)

$$B(r;n,p) = \sum_{k=0}^{r} {}_nC_k\, p^k(1-p)^{n-k}$$

n	r	0.05	0.10	0.15	0.20	0.25	0.30	0.35	0.40	0.45	0.50
2	0	0.9025	0.8100	0.7225	0.6400	0.5625	0.4900	0.4225	0.3600	0.3025	0.2500
	1	0.9975	0.9900	0.9775	0.9600	0.9375	0.9100	0.8775	0.8400	0.7975	0.7500
3	0	0.8574	0.7290	0.6141	0.5120	0.4219	0.3430	0.2746	0.2160	0.1664	0.1250
	1	0.9928	0.9720	0.9392	0.8960	0.8438	0.7840	0.7182	0.6480	0.5748	0.5000
	2	0.9999	0.9990	0.9966	0.9920	0.9844	0.9730	0.9571	0.9360	0.9089	0.8750
4	0	0.8145	0.6561	0.5220	0.4096	0.3164	0.2401	0.1785	0.1296	0.0915	0.0625
	1	0.9860	0.9477	0.8905	0.8192	0.7383	0.6517	0.5630	0.4752	0.3910	0.3125
	2	0.9995	0.9963	0.9880	0.9728	0.9492	0.9163	0.8735	0.8208	0.7585	0.6875
	3	1.0000	0.9999	0.9995	0.9984	0.9961	0.9919	0.9850	0.9744	0.9590	0.9375
5	0	0.7738	0.5905	0.4437	0.3277	0.2373	0.1681	0.1160	0.0778	0.0503	0.0312
	1	0.9774	0.9185	0.8352	0.7373	0.6328	0.5282	0.4284	0.3370	0.2562	0.1875
	2	0.9988	0.9914	0.9734	0.9421	0.8965	0.8369	0.7648	0.6826	0.5931	0.5000
	3	1.0000	0.9995	0.9978	0.9933	0.9844	0.9692	0.9460	0.9130	0.8688	0.8125
	4	1.0000	1.0000	0.9999	0.9997	0.9990	0.9976	0.9947	0.9898	0.9815	0.9688
6	0	0.7351	0.5314	0.3771	0.2621	0.1780	0.1176	0.0754	0.0467	0.0277	0.0156
	1	0.9672	0.8857	0.7765	0.6554	0.5339	0.4202	0.3191	0.2333	0.1636	0.1094
	2	0.9978	0.9842	0.9527	0.9011	0.8306	0.7443	0.6471	0.5443	0.4415	0.3438
	3	0.9999	0.9987	0.9941	0.9830	0.9624	0.9295	0.8826	0.8208	0.7447	0.6562
	4	1.0000	0.9999	0.9996	0.9984	0.9954	0.9891	0.9777	0.9590	0.9308	0.8906
	5	1.0000	1.0000	1.0000	0.9999	0.9998	0.9993	0.9982	0.9959	0.9917	0.9844
7	0	0.6983	0.4783	0.3206	0.2097	0.1335	0.0824	0.0490	0.0280	0.0152	0.0078
	1	0.9556	0.8503	0.7166	0.5767	0.4449	0.3294	0.2338	0.1586	0.1024	0.0625
	2	0.9962	0.9743	0.9262	0.8520	0.7564	0.6471	0.5323	0.4199	0.3164	0.2266
	3	0.9998	0.9973	0.9879	0.9667	0.9294	0.8740	0.8002	0.7102	0.6083	0.5000
	4	1.0000	0.9998	0.9988	0.9953	0.9871	0.9712	0.9444	0.9037	0.8471	0.7734
	5	1.0000	1.0000	0.9999	0.9996	0.9987	0.9962	0.9910	0.9812	0.9643	0.9375
	6	1.0000	1.0000	1.0000	1.0000	0.9999	0.9998	0.9994	0.9984	0.9963	0.9922
8	0	0.6634	0.4305	0.2725	0.1678	0.1001	0.0576	0.0319	0.0168	0.0084	0.0039
	1	0.9428	0.8131	0.6572	0.5033	0.3671	0.2553	0.1691	0.1064	0.0632	0.0352
	2	0.9942	0.9619	0.8948	0.7969	0.6785	0.5518	0.4278	0.3154	0.2201	0.1445
	3	0.9996	0.9950	0.9786	0.9437	0.8862	0.8059	0.7064	0.5941	0.4770	0.3633
	4	1.0000	0.9996	0.9971	0.9896	0.9727	0.9420	0.8939	0.8263	0.7396	0.6367
	5	1.0000	1.0000	0.9998	0.9988	0.9958	0.9887	0.9747	0.9502	0.9115	0.8555
	6	1.0000	1.0000	1.0000	0.9999	0.9996	0.9987	0.9964	0.9915	0.9819	0.9648
	7	1.0000	1.0000	1.0000	1.0000	1.0000	0.9999	0.9998	0.9993	0.9983	0.9961
9	0	0.6302	0.3874	0.2316	0.1342	0.0751	0.0404	0.0207	0.0101	0.0046	0.0020
	1	0.9288	0.7748	0.5995	0.4362	0.3003	0.1960	0.1211	0.0705	0.0385	0.0195
	2	0.9916	0.9470	0.8591	0.7382	0.6007	0.4628	0.3373	0.2318	0.1495	0.0898
	3	0.9994	0.9917	0.9661	0.9144	0.8343	0.7297	0.6089	0.4826	0.3614	0.2539
	4	1.0000	0.9991	0.9944	0.9804	0.9511	0.9012	0.8283	0.7334	0.6214	0.5000
	5	1.0000	0.9999	0.9994	0.9969	0.9900	0.9747	0.9464	0.9006	0.8342	0.7461
	6	1.0000	1.0000	1.0000	0.9997	0.9987	0.9957	0.9888	0.9750	0.9502	0.9102
	7	1.0000	1.0000	1.0000	1.0000	0.9999	0.9996	0.9986	0.9962	0.9909	0.9805
	8	1.0000	1.0000	1.0000	1.0000	1.0000	1.0000	0.9999	0.9997	0.9992	0.9980

TABLE A-1 (continued)

n	r	0.05	0.10	0.15	0.20	0.25	0.30	0.35	0.40	0.45	0.50
10	0	0.5987	0.3487	0.1969	0.1074	0.0563	0.0282	0.0135	0.0060	0.0025	0.0010
	1	0.9139	0.7361	0.5443	0.3758	0.2440	0.1493	0.0860	0.0464	0.0232	0.0107
	2	0.9885	0.9298	0.8202	0.6778	0.5256	0.3828	0.2616	0.1673	0.0996	0.0547
	3	0.9990	0.9872	0.9500	0.8791	0.7759	0.6496	0.5138	0.3823	0.2660	0.1719
	4	0.9999	0.9984	0.9901	0.9672	0.9219	0.8497	0.7515	0.6331	0.5044	0.3770
	5	1.0000	0.9999	0.9986	0.9936	0.9803	0.9527	0.9051	0.8338	0.7384	0.6230
	6	1.0000	1.0000	0.9999	0.9991	0.9965	0.9894	0.9740	0.9452	0.8980	0.8281
	7	1.0000	1.0000	1.0000	0.9999	0.9996	0.9984	0.9952	0.9877	0.9726	0.9453
	8	1.0000	1.0000	1.0000	1.0000	1.0000	0.9999	0.9995	0.9983	0.9955	0.9893
	9	1.0000	1.0000	1.0000	1.0000	1.0000	1.0000	1.0000	0.9999	0.9997	0.9990
11	0	0.5688	0.3138	0.1673	0.0859	0.0422	0.0198	0.0088	0.0036	0.0014	0.0005
	1	0.8981	0.6974	0.4922	0.3221	0.1971	0.1130	0.0606	0.0302	0.0139	0.0059
	2	0.9848	0.9104	0.7788	0.6174	0.4552	0.3127	0.2001	0.1189	0.0652	0.0327
	3	0.9984	0.9815	0.9306	0.8389	0.7133	0.5696	0.4256	0.2963	0.1911	0.1133
	4	0.9999	0.9972	0.9841	0.9496	0.8854	0.7897	0.6683	0.5328	0.3971	0.2744
	5	1.0000	0.9997	0.9973	0.9883	0.9657	0.9218	0.8513	0.7535	0.6331	0.5000
	6	1.0000	1.0000	0.9997	0.9980	0.9924	0.9784	0.9499	0.9006	0.8262	0.7256
	7	1.0000	1.0000	1.0000	0.9998	0.9988	0.9957	0.9878	0.9707	0.9390	0.8867
	8	1.0000	1.0000	1.0000	1.0000	0.9999	0.9994	0.9980	0.9941	0.9852	0.9673
	9	1.0000	1.0000	1.0000	1.0000	1.0000	1.0000	0.9998	0.9993	0.9978	0.9941
	10	1.0000	1.0000	1.0000	1.0000	1.0000	1.0000	1.0000	1.0000	0.9998	0.9995
12	0	0.5404	0.2824	0.1422	0.0687	0.0317	0.0138	0.0057	0.0022	0.0008	0.0002
	1	0.8816	0.6590	0.4435	0.2749	0.1584	0.0850	0.0424	0.0196	0.0083	0.0032
	2	0.9804	0.8891	0.7358	0.5583	0.3907	0.2528	0.1513	0.0834	0.0421	0.0193
	3	0.9978	0.9744	0.9078	0.7946	0.6488	0.4925	0.3467	0.2253	0.1345	0.0730
	4	0.9998	0.9957	0.9761	0.9274	0.8424	0.7237	0.5833	0.4382	0.3044	0.1938
	5	1.0000	0.9995	0.9954	0.9806	0.9456	0.8822	0.7873	0.6652	0.5269	0.3872
	6	1.0000	0.9999	0.9993	0.9961	0.9857	0.9614	0.9154	0.8418	0.7393	0.6128
	7	1.0000	1.0000	0.9999	0.9994	0.9972	0.9905	0.9745	0.9427	0.8883	0.8062
	8	1.0000	1.0000	1.0000	0.9999	0.9996	0.9983	0.9944	0.9847	0.9644	0.9270
	9	1.0000	1.0000	1.0000	1.0000	1.0000	0.9998	0.9992	0.9972	0.9921	0.9807
	10	1.0000	1.0000	1.0000	1.0000	1.0000	1.0000	0.9999	0.9997	0.9989	0.9968
	11	1.0000	1.0000	1.0000	1.0000	1.0000	1.0000	1.0000	1.0000	0.9999	0.9998
13	0	0.5133	0.2542	0.1209	0.0550	0.0238	0.0097	0.0037	0.0013	0.0004	0.0001
	1	0.8646	0.6213	0.3983	0.2336	0.1267	0.0637	0.0296	0.0126	0.0049	0.0017
	2	0.9755	0.8661	0.6920	0.5017	0.3326	0.2025	0.1132	0.0579	0.0269	0.0112
	3	0.9969	0.9658	0.8820	0.7473	0.5843	0.4206	0.2783	0.1686	0.0929	0.0461
	4	0.9997	0.9935	0.9658	0.9009	0.7940	0.6543	0.5005	0.3530	0.2279	0.1334
	5	1.0000	0.9991	0.9925	0.9700	0.9198	0.8346	0.7159	0.5744	0.4268	0.2905
	6	1.0000	0.9999	0.9987	0.9930	0.9757	0.9376	0.8705	0.7712	0.6437	0.5000
	7	1.0000	1.0000	0.9998	0.9988	0.9944	0.9818	0.9538	0.9023	0.8212	0.7095
	8	1.0000	1.0000	1.0000	0.9998	0.9990	0.9960	0.9874	0.9679	0 9302	0.8666
	9	1.0000	1.0000	1.0000	1.0000	0.9999	0.9993	0.9975	0.9922	0.9797	0.9539
	10	1.0000	1.0000	1.0000	1.0000	1.0000	0.9999	0.9997	0.9987	0.9959	0.9888
	11	1.0000	1.0000	1.0000	1.0000	1.0000	1.0000	1.0000	0.9999	0.9995	0.9983
	12	1.0000	1.0000	1.0000	1.0000	1.0000	1.0000	1.0000	1.0000	1.0000	0.9999
14	0	0.4877	0.2288	0.1028	0.0440	0.0178	0.0068	0.0024	0.0008	0.0002	0.0001
	1	0.8470	0.5846	0.3567	0.1979	0.1010	0.0475	0.0205	0.0081	0.0029	0.0009

TABLE A-1 (continued)

n	r	0.05	0.10	0.15	0.20	0.25	0.30	0.35	0.40	0.45	0.50
14	2	0.9699	0.8416	0.6479	0.4481	0.2811	0.1608	0.0839	0.0398	0.0170	0.0065
	3	0.9958	0.9559	0.8535	0.6982	0.5213	0.3552	0.2205	0.1243	0.0632	0.0287
	4	0.9996	0.9908	0.9533	0.8702	0.7415	0.5842	0.4227	0.2793	0.1672	0.0898
	5	1.0000	0.9985	0.9885	0.9561	0.8883	0.7805	0.6405	0.4859	0.3373	0.2120
	6	1.0000	0.9998	0.9978	0.9884	0.9617	0.9067	0.8164	0.6925	0.5461	0.3953
	7	1.0000	1.0000	0.9997	0.9976	0.9897	0.9685	0.9247	0.8499	0.7414	0.6047
	8	1.0000	1.0000	1.0000	0.9996	0.9978	0.9917	0.9757	0.9417	0.8811	0.7880
	9	1.0000	1.0000	1.0000	1.0000	0.9997	0.9983	0.9940	0.9825	0.9574	0.9102
	10	1.0000	1.0000	1.0000	1.0000	1.0000	0.9998	0.9989	0.9961	0.9886	0.9713
	11	1.0000	1.0000	1.0000	1.0000	1.0000	1.0000	0.9999	0.9994	0.9978	0.9935
	12	1.0000	1.0000	1.0000	1.0000	1.0000	1.0000	1.0000	0.9999	0.9997	0.9991
	13	1.0000	1.0000	1.0000	1.0000	1.0000	1.0000	1.0000	1.0000	1.0000	0.9999
15	0	0.4633	0.2059	0.0874	0.0352	0.0134	0.0047	0.0016	0.0005	0.0001	0.0000
	1	0.8290	0.5490	0.3186	0.1671	0.0802	0.0353	0.0142	0.0052	0.0017	0.0005
	2	0.9638	0.8159	0.6042	0.3980	0.2361	0.1268	0.0617	0.0271	0.0107	0.0037
	3	0.9945	0.9444	0.8227	0.6482	0.4613	0.2969	0.1727	0.0905	0.0424	0.0176
	4	0.9994	0.9873	0.9383	0.8358	0.6865	0.5155	0.3519	0.2173	0.1204	0.0592
	5	0.9999	0.9978	0.9832	0.9389	0.8516	0.7216	0.5643	0.4032	0.2608	0.1509
	6	1.0000	0.9997	0.9964	0.9819	0.9434	0.8689	0.7548	0.6098	0.4522	0.3036
	7	1.0000	1.0000	0.9996	0.9958	0.9827	0.9500	0.8868	0.7869	0.6535	0.5000
	8	1.0000	1.0000	0.9999	0.9992	0.9958	0.9848	0.9578	0.9050	0.8182	0.6964
	9	1.0000	1.0000	1.0000	0.9999	0.9992	0.9963	0.9876	0.9662	0.9231	0.8491
	10	1.0000	1.0000	1.0000	1.0000	0.9999	0.9993	0.9972	0.9907	0.9745	0.9408
	11	1.0000	1.0000	1.0000	1.0000	1.0000	0.9999	0.9995	0.9981	0.9937	0.9824
	12	1.0000	1.0000	1.0000	1.0000	1.0000	1.0000	0.9999	0.9997	0.9989	0.9963
	13	1.0000	1.0000	1.0000	1.0000	1.0000	1.0000	1.0000	1.0000	0.9999	0.9995
	14	1.0000	1.0000	1.0000	1.0000	1.0000	1.0000	1.0000	1.0000	1.0000	1.0000
16	0	0.4401	0.1853	0.0743	0.0281	0.0100	0.0033	0.0010	0.0003	0.0001	0.0000
	1	0.8108	0.5147	0.2839	0.1407	0.0635	0.0261	0.0098	0.0033	0.0010	0.0003
	2	0.9571	0.7892	0.5614	0.3518	0.1971	0.0994	0.0451	0.0183	0.0066	0.0021
	3	0.9930	0.9316	0.7899	0.5981	0.4050	0.2459	0.1339	0.0651	0.0281	0.0106
	4	0.9991	0.9830	0.9209	0.7982	0.6302	0.4499	0.2892	0.1666	0.0853	0.0384
	5	0.9999	0.9967	0.9765	0.9183	0.8103	0.6598	0.4900	0.3288	0.1976	0.1051
	6	1.0000	0.9995	0.9944	0.9733	0.9204	0.8247	0.6881	0.5272	0.3660	0.2272
	7	1.0000	0.9999	0.9989	0.9930	0.9729	0.9256	0.8406	0.7161	0.5629	0.4018
	8	1.0000	1.0000	0.9998	0.9985	0.9925	0.9743	0.9329	0.8577	0.7441	0.5982
	9	1.0000	1.0000	1.0000	0.9998	0.9984	0.9929	0.9771	0.9417	0.8759	0.7728
	10	1.0000	1.0000	1.0000	1.0000	0.9997	0.9984	0.9938	0.9809	0.9514	0.8949
	11	1.0000	1.0000	1.0000	1.0000	1.0000	0.9997	0.9987	0.9951	0.9851	0.9616
	12	1.0000	1.0000	1.0000	1.0000	1.0000	1.0000	0.9998	0.9991	0.9965	0.9894
	13	1.0000	1.0000	1.0000	1.0000	1.0000	1.0000	1.0000	0.9999	0.9994	0.9979
	14	1.0000	1.0000	1.0000	1.0000	1.0000	1.0000	1.0000	1.0000	1.0000	0.9997
	15	1.0000	1.0000	1.0000	1.0000	1.0000	1.0000	1.0000	1.0000	1.0000	1.0000
17	0	0.4181	0.1668	0.0631	0.0225	0.0075	0.0023	0.0007	0.0002	0.0000	0.0000
	1	0.7922	0.4818	0.2525	0.1182	0.0501	0.0193	0.0067	0.0021	0.0006	0.0001
	2	0.9497	0.7618	0.5198	0.3096	0.1637	0.0774	0.0327	0.0123	0.0041	0.0012
	3	0.9912	0.9174	0.7556	0.5489	0.3530	0.2019	0.1028	0.0464	0.0184	0.0064
	4	0.9988	0.9779	0.9013	0.7582	0.5739	0.3887	0.2348	0.1260	0.0596	0.0245

TABLE A-1 (continued)

n	r	0.05	0.10	0.15	0.20	0.25	0.30	0.35	0.40	0.45	0.50
17	5	0.9999	0.9953	0.9681	0.8943	0.7653	0.5968	0.4197	0.2639	0.1471	0.0717
	6	1.0000	0.9992	0.9917	0.9623	0.8929	0.7752	0.6188	0.4478	0.2902	0.1662
	7	1.0000	0.9999	0.9983	0.9891	0.9598	0.8954	0.7872	0.6405	0.4743	0.3145
	8	1.0000	1.0000	0.9997	0.9974	0.9876	0.9597	0.9006	0.8011	0.6626	0.5000
	9	1.0000	1.0000	1.0000	0.9995	0.9969	0.9873	0.9617	0.9081	0.8166	0.6855
	10	1.0000	1.0000	1.0000	0.9999	0.9994	0.9968	0.9880	0.9652	0.9174	0.8338
	11	1.0000	1.0000	1.0000	1.0000	0.9999	0.9993	0.9970	0.9894	0.9699	0.9283
	12	1.0000	1.0000	1.0000	1.0000	1.0000	0.9999	0.9994	0.9975	0.9914	0.9755
	13	1.0000	1.0000	1.0000	1.0000	1.0000	1.0000	0.9999	0.9995	0.9981	0.9936
	14	1.0000	1.0000	1.0000	1.0000	1.0000	1.0000	1.0000	0.9999	0.9997	0.9988
	15	1.0000	1.0000	1.0000	1.0000	1.0000	1.0000	1.0000	1.0000	1.0000	0.9999
	16	1.0000	1.0000	1.0000	1.0000	1.0000	1.0000	1.0000	1.0000	1.0000	1.0000
18	0	0.3972	0.1501	0.0536	0.0180	0.0056	0.0016	0.0004	0.0001	0.0000	0.0000
	1	0.7735	0.4503	0.2241	0.0991	0.0395	0.0142	0.0046	0.0013	0.0003	0.0001
	2	0.9419	0.7338	0.4797	0.2713	0.1353	0.0600	0.0236	0.0082	0.0025	0.0007
	3	0.9891	0.9018	0.7202	0.5010	0.3057	0.1646	0.0783	0.0328	0.0120	0.0038
	4	0.9985	0.9718	0.8794	0.7164	0.5187	0.3327	0.1886	0.0942	0.0411	0.0154
	5	0.9998	0.9936	0.9581	0.8671	0.7175	0.5344	0.3550	0.2088	0.1077	0.0481
	6	1.0000	0.9988	0.9882	0.9487	0.8610	0.7217	0.5491	0.3743	0.2258	0.1189
	7	1.0000	0.9998	0.9973	0.9837	0.9431	0.8593	0.7283	0.5634	0.3915	0.2403
	8	1.0000	1.0000	0.9995	0.9957	0.9807	0.9404	0.8609	0.7368	0.5778	0.4073
	9	1.0000	1.0000	0.9999	0.9991	0.9946	0.9790	0.9403	0.8653	0.7473	0.5927
	10	1.0000	1.0000	1.0000	0.9998	0.9988	0.9939	0.9788	0.9424	0.8720	0.7597
	11	1.0000	1.0000	1.0000	1.0000	0.9998	0.9986	0.9938	0.9797	0.9463	0.8811
	12	1.0000	1.0000	1.0000	1.0000	1.0000	0.9997	0.9986	0.9942	0.9817	0.9519
	13	1.0000	1.0000	1.0000	1.0000	1.0000	1.0000	0.9997	0.9987	0.9951	0.9846
	14	1.0000	1.0000	1.0000	1.0000	1.0000	1.0000	1.0000	0.9998	0.9990	0.9962
	15	1.0000	1.0000	1.0000	1.0000	1.0000	1.0000	1.0000	1.0000	0.9999	0.9993
	16	1.0000	1.0000	1.0000	1.0000	1.0000	1.0000	1.0000	1.0000	1.0000	0.9999
19	0	0.3774	0.1351	0.0456	0.0144	0.0042	0.0011	0.0003	0.0001	0.0000	0.0000
	1	0.7547	0.4203	0.1985	0.0829	0.0310	0.0104	0.0031	0.0008	0.0002	0.0000
	2	0.9335	0.7054	0.4413	0.2369	0.1113	0.0462	0.0170	0.0055	0.0015	0.0004
	3	0.9868	0.8850	0.6841	0.4551	0.2630	0.1332	0.0591	0.0230	0.0077	0.0022
	4	0.9980	0.9648	0.8556	0.6733	0.4654	0.2822	0.1500	0.0696	0.0280	0.0096
	5	0.9998	0.9914	0.9463	0.8369	0.6678	0.4739	0.2968	0.1629	0.0777	0.0318
	6	1.0000	0.9983	0.9837	0.9324	0.8251	0.6655	0.4812	0.3081	0.1727	0.0835
	7	1.0000	0.9997	0.9959	0.9767	0.9225	0.8180	0.6656	0.4878	0.3169	0.1796
	8	1.0000	1.0000	0.9992	0.9933	0.9713	0.9161	0.8145	0.6675	0.4940	0.3238
	9	1.0000	1.0000	0.9999	0.9984	0.9911	0.9674	0.9125	0.8139	0.6710	0.5000
	10	1.0000	1.0000	1.0000	0.9997	0.9977	0.9895	0.9653	0.9115	0.8159	0.6762
	11	1.0000	1.0000	1.0000	1.0000	0.9995	0.9972	0.9886	0.9648	0.9129	0.8204
	12	1.0000	1.0000	1.0000	1.0000	0.9999	0.9994	0.9969	0.9884	0.9658	0.9165
	13	1.0000	1.0000	1.0000	1.0000	1.0000	0.9999	0.9993	0.9969	0.9891	0.9682
	14	1.0000	1.0000	1.0000	1.0000	1.0000	1.0000	0.9999	0.9994	0.9972	0.9904
	15	1.0000	1.0000	1.0000	1.0000	1.0000	1.0000	1.0000	0.9999	0.9995	0.9978
	16	1.0000	1.0000	1.0000	1.0000	1.0000	1.0000	1.0000	1.0000	0.9999	0.9996
	17	1.0000	1.0000	1.0000	1.0000	1.0000	1.0000	1.0000	1.0000	1.0000	1.0000

TABLE A-1 (continued)

n	r	0.05	0.10	0.15	0.20	0.25	0.30	0.35	0.40	0.45	0.50
20	0	0.3585	0.1216	0.0388	0.0115	0.0032	0.0008	0.0002	0.0000	0.0000	0.0000
	1	0.7358	0.3917	0.1756	0.0692	0.0243	0.0076	0.0021	0.0005	0.0001	0.0000
	2	0.9245	0.6769	0.4049	0.2061	0.0913	0.0355	0.0121	0.0036	0.0009	0.0002
	3	0.9841	0.8670	0.6477	0.4114	0 2252	0.1071	0.0444	0.0160	0.0049	0.0013
	4	0.9974	0.9568	0.8298	0.6296	0.4148	0.2375	0.1182	0.0510	0.0189	0.0059
	5	0.9997	0.9887	0.9327	0.8042	0.6172	0.4164	0.2454	0.1256	0.0553	0.0207
	6	1.0000	0.9976	0.9781	0.9133	0.7858	0.6080	0.4166	0.2500	0.1299	0.0577
	7	1.0000	0.9996	0.9941	0.9679	0.8982	0.7723	0.6010	0.4159	0.2520	0.1316
	8	1.0000	0.9999	0.9987	0.9900	0.9591	0.8867	0.7624	0.5956	0.4143	0.2517
	9	1.0000	1.0000	0.9998	0.9974	0.9861	0 9520	0.8782	0.7553	0.5914	0.4119
	10	1.0000	1.0000	1.0000	0.9994	0.9961	0.9829	0.9468	0.8725	0.7507	0.5881
	11	1.0000	1.0000	1.0000	0.9999	0.9991	0.9949	0.9804	0.9435	0.8692	0.7483
	12	1.0000	1.0000	1.0000	1.0000	0.9998	0.9987	0.9940	0.9790	0.9420	0.8684
	13	1.0000	1.0000	1.0000	1.0000	1.0000	0.9997	0.9985	0.9935	0.9786	0.9423
	14	1.0000	1.0000	1.0000	1.0000	1.0000	1.0000	0.9997	0.9984	0.9936	0.9793
	15	1.0000	1.0000	1.0000	1.0000	1.0000	1.0000	1.0000	0.9997	0.9985	0.9941
	16	1.0000	1.0000	1.0000	1.0000	1.0000	1.0000	1.0000	1.0000	0.9997	0.9987
	17	1.0000	1.0000	1.0000	1.0000	1.0000	1.0000	1.0000	1.0000	1.0000	0.9998
	18	1.0000	1.0000	1.0000	1.0000	1.0000	1.0000	1.0000	1.0000	1.0000	1.0000

Source: Irwin Miller and John E. Freund, *Probability and Statistics for Engineers,* © 1965, pp. 388–392. Reprinted by permission of Prentice-Hall, Inc., Englewood Cliffs, N.J.

TABLE A-2. Poisson Distribution (Cumulative)

$$\Pr(x \text{ or less}) = \sum_{k=0}^{x} \frac{e^{-m}m^{x}}{x!}$$

					m				
x	.1	.2	.3	.4	.5	.6	.7	.8	x
0	.905	.819	.741	.670	.607	.549	.497	.449	0
1	.995	.982	.963	.938	.910	.878	.844	.809	1
2	1.000	.999	.996	.992	.986	.977	.966	.953	2
3		1.000	1.000	.999	.998	.997	.994	.991	3
4				1.000	1.000	1.000	.999	.999	4
5							1.000	1.000	5

					m				
x	.9	1.0	1.5	2.0	2.5	3.0	3.5	4.0	x
0	.407	.368	.223	.135	.082	.050	.030	.018	0
1	.772	.736	.558	.406	.287	.199	.136	.092	1
2	.937	.920	.809	.677	.544	.423	.321	.238	2
3	.987	.981	.934	.857	.758	.647	.537	.433	3
4	.998	.996	.981	.947	.891	.815	.725	.629	4
5	1.000	.999	.996	.983	.958	.916	.858	.785	5
6		1.000	.999	.995	.986	.966	.935	.889	6
7			1.000	.999	.996	.988	.973	.949	7
8				1.000	.999	.996	.990	.979	8
9					1.000	.999	.997	.992	9
10						1.000	.999	.997	10
11							1.000	.999	11
12								1.000	12

TABLE A-2 (continued)

					m				
x	4.5	5.0	5.5	6.0	6.5	7.0	7.5	8.0	x
0	.011	.007	.004	.002	.002	.001	.001	.000	0
1	.061	.040	.027	.017	.011	.007	.005	.003	1
2	.174	.125	.088	.062	.043	.030	.020	.014	2
3	.342	.265	.202	.151	.112	.082	.059	.042	3
4	.532	.440	.358	.285	.224	.173	.132	.100	4
5	.703	.616	.529	.446	.369	.301	.241	.191	5
6	.831	.762	.686	.606	.527	.450	.378	.313	6
7	.913	.867	.809	.744	.673	.599	.525	.453	7
8	.960	.932	.894	.847	.792	.729	.662	.593	8
9	.983	.968	.946	.916	.877	.830	.776	.717	9
10	.993	.986	.975	.957	.933	.901	.862	.816	10
11	.998	.995	.989	.980	.966	.947	.921	.888	11
12	.999	.998	.996	.991	.984	.973	.957	.936	12
13	1.000	.999	.998	.996	.993	.987	.978	.966	13
14		1.000	.999	.999	.997	.994	.990	.983	14
15			1.000	.999	.999	.998	.995	.992	15
16				1.000	1.000	.999	.998	.996	16
17						1.000	.999	.998	17
18							1.000	.999	18
19								1.000	19

			m		
x	8.5	9.0	9.5	10.0	x
0	.000	.000	.000	.000	0
1	.002	.001	.001	.000	1
2	.009	.006	.004	.003	2
3	.030	.021	.015	.010	3
4	.074	.055	.040	.029	4
5	.150	.116	.089	.067	5
6	.256	.207	.165	.130	6
7	.386	.324	.269	.220	7
8	.523	.456	.392	.333	8
9	.653	.587	.522	.458	9
10	.763	.706	.645	.583	10
11	.849	.803	.752	.697	11
12	.909	.876	.836	.792	12
13	.949	.926	.898	.864	13
14	.973	.959	.940	.917	14
15	.986	.978	.967	.951	15
16	.993	.989	.982	.973	16
17	.997	.995	.991	.986	17
18	.999	.998	.996	.993	18
19	.999	.999	.998	.997	19
20	1.000	1.000	.999	.998	20
21			1.000	.999	21
22				1.000	22

TABLE A-3. Normal Distribution (Area under the Standard Normal Curve from 0 to z)

z	0.00	0.01	0.02	0.03	0.04	0.05	0.06	0.07	0.08	0.09
0.0	0.0000	0.0040	0.0080	0.0120	0.0160	0.0199	0.0239	0.0279	0.0319	0.0359
0.1	.0398	.0438	.0478	.0517	.0557	.0596	.0636	.0675	.0714	.0753
0.2	.0793	.0832	.0871	.0910	.0948	.0987	.1026	.1064	.1103	.1141
0.3	.1179	.1217	.1255	.1293	.1331	.1368	.1406	.1443	.1480	.1517
0.4	.1554	.1591	.1628	.1664	.1700	.1736	.1772	.1808	.1844	.1879
0.5	.1915	.1950	.1985	.2019	.2054	.2088	.2123	.2157	.2190	.2224
0.6	.2257	.2291	.2324	.2357	.2389	.2422	.2454	.2486	.2517	.2549
0.7	.2580	.2611	.2642	.2673	.2704	.2734	.2764	.2794	.2823	.2852
0.8	.2881	.2910	.2939	.2967	.2995	.3023	.3051	.3078	.3106	.3133
0.9	.3159	.3186	.3212	.3238	.3264	.3289	.3315	.3340	.3365	.3389
1.0	.3413	.3438	.3461	.3485	.3508	.3531	.3554	.3577	.3599	.3621
1.1	.3643	.3665	.3686	.3708	.3729	.3749	.3770	.3790	.3810	.3830
1.2	.3849	.3869	.3888	.3907	.3925	.3944	.3962	.3980	.3997	.4015
1.3	.4032	.4049	.4066	.4082	.4099	.4115	.4131	.4147	.4162	.4177
1.4	.4192	.4207	.4222	.4236	.4251	.4265	.4279	.4292	.4306	.4319
1.5	.4332	.4345	.4357	.4370	.4382	.4394	.4406	.4418	.4429	.4441
1.6	.4452	.4463	.4474	.4484	.4495	.4505	.4515	.4525	.4535	.4545
1.7	.4554	.4564	.4573	.4582	.4591	.4599	.4608	.4616	.4625	.4633
1.8	.4641	.4649	.4656	.4664	.4671	.4678	.4686	.4693	.4699	.4706
1.9	.4713	.4719	.4726	.4732	.4738	.4744	.4750	.4756	.4761	.4767
2.0	.4772	.4778	.4783	.4788	.4793	.4798	.4803	.4808	.4812	.4817
2.1	.4821	.4826	.4830	.4834	.4838	.4842	.4846	.4850	.4854	.4857
2.2	.4861	.4864	.4868	.4871	.4875	.4878	.4881	.4884	.4887	.4890
2.3	.4893	.4896	.4898	.4901	.4904	.4906	.4909	.4911	.4913	.4916
2.4	.4918	.4920	.4922	.4925	.4927	.4929	.4931	.4932	.4934	.4936
2.5	.4938	.4940	.4941	.4943	.4945	.4946	.4948	.4949	.4951	.4952
2.6	.4953	.4955	.4956	.4957	.4959	.4960	.4961	.4962	.4963	.4964
2.7	.4965	.4966	.4967	.4968	.4969	.4970	.4971	.4972	.4973	.4974
2.8	.4974	.4975	.4976	.4977	.4977	.4978	.4979	.4979	.4980	.4981
2.9	.4981	.4982	.4982	.4983	.4984	.4984	.4985	.4985	.4986	.4986
3.0	.4987	.4987	.4987	.4988	.4988	.4989	.4989	.4989	.4990	.4990
3.1	.4990	.4991	.4991	.4991	.4992	.4992	.4992	.4992	.4993	.4993
3.2	.4993	.4993	.4994	.4994	.4994	.4994	.4994	.4995	.4995	.4995
3.3	.4995	.4995	.4995	.4996	.4996	.4996	.4996	.4996	.4996	.4997
3.4	.4997	.4997	.4997	.4997	.4997	.4997	.4997	.4997	.4997	.4998
3.6	.4998	.4998	.4999	.4999	.4999	.4999	.4999	.4999	.4999	.4999
3.9	.5000									

Source: Reprinted by permission from *Statistical Methods* by George W. Snedecor and William G. Cochran, sixth edition, © 1967 by Iowa State University Press, Ames, Iowa.

TABLE A-4. t Distribution

		Level of Significance (α)					
	One Tail	.15	.10	.05	.025	.010	.005
df	Two Tails	.30	.20	.10	.050	.020	.010
1		1.963	3.078	6.314	12.706	31.821	63.657
2		1.386	1.886	2.920	4.303	6.965	9.925
3		1.250	1.638	2.353	3.182	4.541	5.841
4		1.190	1.533	2.132	2.776	3.747	4.604
5		1.156	1.476	2.015	2.571	3.365	4.032
6		1.134	1.440	1.943	2.447	3.143	3.707
7		1.119	1.415	1.895	2.365	2.998	3.499
8		1.108	1.397	1.860	2.306	2.896	3.355
9		1.100	1.383	1.833	2.262	2.821	3.250
10		1.093	1.372	1.812	2.228	2.764	3.169
11		1.088	1.363	1.796	2.201	2.718	3.106
12		1.083	1.356	1.782	2.179	2.681	3.055
13		1.079	1.350	1.771	2.160	2.650	3.012
14		1.076	1.345	1.761	2.145	2.624	2.977
15		1.074	1.341	1.753	2.131	2.602	2.947
16		1.071	1.337	1.746	2.120	2.583	2.921
17		1.069	1.333	1.740	2.110	2.567	2.898
18		1.067	1.330	1.734	2.101	2.552	2.878
19		1.066	1.328	1.729	2.093	2.539	2.861
20		1.064	1.325	1.725	2.086	2.528	2.845
21		1.063	1.323	1.721	2.080	2.518	2.831
22		1.061	1.321	1.717	2.074	2.508	2.819
23		1.060	1.319	1.714	2.069	2.500	2.807
24		1.059	1.318	1.711	2.064	2.492	2.797
25		1.058	1.316	1.708	2.060	2.485	2.787
26		1.058	1.315	1.706	2.056	2.479	2.779
27		1.057	1.314	1.703	2.052	2.473	2.771
28		1.056	1.313	1.701	2.048	2.467	2.763
29		1.055	1.311	1.699	2.045	2.462	2.756
30		1.055	1.310	1.697	2.042	2.457	2.750
∞		1.03643	1.28155	1.64485	1.95996	2.32634	2.57582

Source: From *Statistical Methods for Research Workers* (14th Ed.) by Ronald A. Fisher (Copyright © 1972 by Hafner Press). Reprinted by permission of the publisher.

TABLE A-5. χ^2 Distribution

df	Level of Significance (α)						
	.50	.30	.20	.10	.05	.02	.01
1	.455	1.074	1.642	2.706	3.841	5.412	6.635
2	1.386	2.408	3.219	4.605	5.991	7.824	9.210
3	2.366	3.665	4.642	6.251	7.815	9.837	11.345
4	3.357	4.878	5.989	7.779	9.488	11.668	13.277
5	4.351	6.064	7.289	9.236	11.070	13.388	15.086
6	5.348	7.231	8.558	10.645	12.592	15.033	16.812
7	6.346	8.383	9.803	12.017	14.067	16.622	18.475
8	7.344	9.524	11.030	13.362	15.507	18.168	20.090
9	8.343	10.656	12.242	14.684	16.919	19.679	21.666
10	9.342	11.781	13.442	15.987	18.307	21.161	23.209
11	10.341	12.899	14.631	17.275	19.675	22.618	24.725
12	11.340	14.011	15.812	18.549	21.026	24.054	26.217
13	12.340	15.119	16.985	19.812	22.362	25.472	27.688
14	13.339	16.222	18.151	21.064	23.685	26.873	29.141
15	14.339	17.322	19.311	22.307	24.996	28.259	30.578
16	15.338	18.418	20.465	23.542	26.296	29.633	32.000
17	16.338	19.511	21.615	24.769	27.587	30.995	33.409
18	17.338	20.601	22.760	25.989	28.869	32.346	34.805
19	18.338	21.689	23.900	27.204	30.144	33.687	36.191
20	19.337	22.775	25.038	28.412	31.410	35.020	37.566
21	20.337	23.858	26.171	29.615	32.671	36.343	38.932
22	21.337	24.939	27.301	30.813	33.924	37.659	40.289
23	22.337	26.018	28.429	32.007	35.172	38.968	41.638
24	23.337	27.096	29.553	33.196	36.415	40.270	42.980
25	24.337	28.172	30.675	34.382	37.652	41.566	44.314
26	25.336	29.246	31.795	35.563	38.885	42.856	45.642
27	26.336	30.319	32.912	36.741	40.113	44.140	46.963
28	27.336	31.391	34.027	37.916	41.337	45.419	48.278
29	28.336	32.461	35.139	39.087	42.557	46.693	49.588
30	29.336	33.530	36.250	40.256	43.773	47.962	50.892

Source: From *Statistical Methods for Research Workers* (14th Ed.) by Ronald A. Fisher (Copyright © 1972 by Hafner Press). Reprinted by permission of the publisher.

TABLE A-6. F Distribution   α = 5% (roman type) and 1% (boldface type)

f_1 Degrees of Freedom (for greater mean square)

f_2	1	2	3	4	5	6	7	8	9	10	11	12	14	16	20	24	30	40	50	75	100	∞	f_2
1	161 **4,052**	200 **4,999**	216 **5,403**	225 **5,625**	230 **5,764**	234 **5,859**	237 **5,928**	239 **5,981**	241 **6,022**	242 **6,056**	243 **6,082**	244 **6,106**	245 **6,142**	246 **6,169**	248 **6,208**	249 **6,234**	250 **6,261**	251 **6,286**	252 **6,302**	253 **6,323**	253 **6,334**	254 **6,366**	1
2	18.51 **98.49**	19.00 **99.00**	19.16 **99.17**	19.25 **99.25**	19.30 **99.30**	19.33 **99.33**	19.36 **99.36**	19.37 **99.37**	19.38 **99.39**	19.39 **99.40**	19.40 **99.41**	19.41 **99.42**	19.42 **99.43**	19.43 **99.44**	19.44 **99.45**	19.45 **99.46**	19.46 **99.47**	19.47 **99.48**	19.47 **99.48**	19.48 **99.49**	19.49 **99.49**	19.50 **99.50**	2
3	10.13 **34.12**	9.55 **30.82**	9.28 **29.46**	9.12 **28.71**	9.01 **28.24**	8.94 **27.91**	8.88 **27.67**	8.84 **27.49**	8.81 **27.34**	8.78 **27.23**	8.76 **27.13**	8.74 **27.05**	8.71 **26.92**	8.69 **26.83**	8.66 **26.69**	8.64 **26.60**	8.62 **26.50**	8.60 **26.41**	8.58 **26.35**	8.57 **26.27**	8.56 **26.23**	8.53 **26.12**	3
4	7.71 **21.20**	6.94 **18.00**	6.59 **16.69**	6.39 **15.98**	6.26 **15.52**	6.16 **15.21**	6.09 **14.98**	6.04 **14.80**	6.00 **14.66**	5.96 **14.54**	5.93 **14.45**	5.91 **14.37**	5.87 **14.24**	5.84 **14.15**	5.80 **14.02**	5.77 **13.93**	5.74 **13.83**	5.71 **13.74**	5.70 **13.69**	5.68 **13.61**	5.66 **13.57**	5.63 **13.46**	4
5	6.61 **16.26**	5.79 **13.27**	5.41 **12.06**	5.19 **11.39**	5.05 **10.97**	4.95 **10.67**	4.88 **10.45**	4.82 **10.29**	4.78 **10.15**	4.74 **10.05**	4.70 **9.96**	4.68 **9.89**	4.64 **9.77**	4.60 **9.68**	4.56 **9.55**	4.53 **9.47**	4.50 **9.38**	4.46 **9.29**	4.44 **9.24**	4.42 **9.17**	4.40 **9.13**	4.36 **9.02**	5
6	5.99 **13.74**	5.14 **10.92**	4.76 **9.78**	4.53 **9.15**	4.39 **8.75**	4.28 **8.47**	4.21 **8.26**	4.15 **8.10**	4.10 **7.98**	4.06 **7.87**	4.03 **7.79**	4.00 **7.72**	3.96 **7.60**	3.92 **7.52**	3.87 **7.39**	3.84 **7.31**	3.81 **7.23**	3.77 **7.14**	3.75 **7.09**	3.72 **7.02**	3.71 **6.99**	3.67 **6.88**	6
7	5.59 **12.25**	4.74 **9.55**	4.35 **8.45**	4.12 **7.85**	3.97 **7.46**	3.87 **7.19**	3.79 **7.00**	3.73 **6.84**	3.68 **6.71**	3.63 **6.62**	3.60 **6.54**	3.57 **6.47**	3.52 **6.35**	3.49 **6.27**	3.44 **6.15**	3.41 **6.07**	3.38 **5.98**	3.34 **5.90**	3.32 **5.85**	3.29 **5.78**	3.28 **5.75**	3.23 **5.65**	7
8	5.32 **11.26**	4.46 **8.65**	4.07 **7.59**	3.84 **7.01**	3.69 **6.63**	3.58 **6.37**	3.50 **6.19**	3.44 **6.03**	3.39 **5.91**	3.34 **5.82**	3.31 **5.74**	3.28 **5.67**	3.23 **5.56**	3.20 **5.48**	3.15 **5.36**	3.12 **5.28**	3.08 **5.20**	3.05 **5.11**	3.03 **5.06**	3.00 **5.00**	2.98 **4.96**	2.93 **4.86**	8
9	5.12 **10.56**	4.26 **8.02**	3.86 **6.99**	3.63 **6.42**	3.48 **6.06**	3.37 **5.80**	3.29 **5.62**	3.23 **5.47**	3.18 **5.35**	3.13 **5.26**	3.10 **5.18**	3.07 **5.11**	3.02 **5.00**	2.98 **4.92**	2.93 **4.80**	2.90 **4.73**	2.86 **4.64**	2.82 **4.56**	2.80 **4.51**	2.77 **4.45**	2.76 **4.41**	2.71 **4.31**	9
10	4.96 **10.04**	4.10 **7.56**	3.71 **6.55**	3.48 **5.99**	3.33 **5.64**	3.22 **5.39**	3.14 **5.21**	3.07 **5.06**	3.02 **4.95**	2.97 **4.85**	2.94 **4.78**	2.91 **4.71**	2.86 **4.60**	2.82 **4.52**	2.77 **4.41**	2.74 **4.33**	2.70 **4.25**	2.67 **4.17**	2.64 **4.12**	2.61 **4.05**	2.59 **4.01**	2.54 **3.91**	10
11	4.84 **9.65**	3.98 **7.20**	3.59 **6.22**	3.36 **5.67**	3.20 **5.32**	3.09 **5.07**	3.01 **4.88**	2.95 **4.74**	2.90 **4.63**	2.86 **4.54**	2.82 **4.46**	2.79 **4.40**	2.74 **4.29**	2.70 **4.21**	2.65 **4.10**	2.61 **4.02**	2.57 **3.94**	2.53 **3.86**	2.50 **3.80**	2.47 **3.74**	2.45 **3.70**	2.40 **3.60**	11
12	4.75 **9.33**	3.88 **6.93**	3.49 **5.95**	3.26 **5.41**	3.11 **5.06**	3.00 **4.82**	2.92 **4.65**	2.85 **4.50**	2.80 **4.39**	2.76 **4.30**	2.72 **4.22**	2.69 **4.16**	2.64 **4.05**	2.60 **3.98**	2.54 **3.86**	2.50 **3.78**	2.46 **3.70**	2.42 **3.61**	2.40 **3.56**	2.36 **3.49**	2.35 **3.46**	2.30 **3.36**	12
13	4.67 **9.07**	3.80 **6.70**	3.41 **5.74**	3.18 **5.20**	3.02 **4.86**	2.92 **4.62**	2.84 **4.44**	2.77 **4.30**	2.72 **4.19**	2.67 **4.10**	2.63 **4.02**	2.60 **3.96**	2.55 **3.85**	2.51 **3.78**	2.46 **3.67**	2.42 **3.59**	2.38 **3.51**	2.34 **3.42**	2.32 **3.37**	2.28 **3.30**	2.26 **3.27**	2.21 **3.16**	13

TABLE A-6 (continued)

f_1 Degrees of Freedom (for greater mean square)

f_2	1	2	3	4	5	6	7	8	9	10	11	12	14	16	20	24	30	40	50	75	100	∞	f_2
14	4.60 / 8.86	3.74 / 6.51	3.34 / 5.56	3.11 / 5.03	2.96 / 4.69	2.85 / 4.46	2.77 / 4.28	2.70 / 4.14	2.65 / 4.03	2.60 / 3.94	2.56 / 3.86	2.53 / 3.80	2.48 / 3.70	2.44 / 3.62	2.39 / 3.51	2.35 / 3.43	2.31 / 3.34	2.27 / 3.26	2.24 / 3.21	2.21 / 3.14	2.19 / 3.11	2.13 / 3.00	14
15	4.54 / 8.68	3.68 / 6.36	3.29 / 5.42	3.06 / 4.89	2.90 / 4.56	2.79 / 4.32	2.70 / 4.14	2.64 / 4.00	2.59 / 3.89	2.55 / 3.80	2.51 / 3.73	2.48 / 3.67	2.43 / 3.56	2.39 / 3.48	2.33 / 3.36	2.29 / 3.29	2.25 / 3.20	2.21 / 3.12	2.18 / 3.07	2.15 / 3.00	2.12 / 2.97	2.07 / 2.87	15
16	4.49 / 8.53	3.63 / 6.23	3.24 / 5.29	3.01 / 4.77	2.85 / 4.44	2.74 / 4.20	2.66 / 4.03	2.59 / 3.89	2.54 / 3.78	2.49 / 3.69	2.45 / 3.61	2.42 / 3.55	2.37 / 3.45	2.33 / 3.37	2.28 / 3.25	2.24 / 3.18	2.20 / 3.10	2.16 / 3.01	2.13 / 2.96	2.09 / 2.98	2.07 / 2.86	2.01 / 2.75	16
17	4.45 / 8.40	3.59 / 6.11	3.20 / 5.18	2.96 / 4.67	2.81 / 4.34	2.70 / 4.10	2.62 / 3.93	2.55 / 3.79	2.50 / 3.68	2.45 / 3.59	2.41 / 3.52	2.38 / 3.45	2.33 / 3.35	2.29 / 3.27	2.23 / 3.16	2.19 / 3.08	2.15 / 3.00	2.11 / 2.92	2.08 / 2.86	2.04 / 2.79	2.02 / 2.76	1.96 / 2.65	17
18	4.41 / 8.28	3.55 / 6.01	3.16 / 5.09	2.93 / 4.58	2.77 / 4.25	2.66 / 4.01	2.58 / 3.85	2.51 / 3.71	2.46 / 3.60	2.41 / 3.51	2.37 / 3.44	2.34 / 3.37	2.29 / 3.27	2.25 / 3.19	2.19 / 3.07	2.15 / 3.00	2.11 / 2.91	2.07 / 2.83	2.04 / 2.78	2.00 / 2.71	1.98 / 2.68	1.92 / 2.57	18
19	4.38 / 8.18	3.52 / 5.93	3.13 / 5.01	2.90 / 4.50	2.74 / 4.17	2.63 / 3.94	2.55 / 3.77	2.48 / 3.63	2.43 / 3.52	2.38 / 3.43	2.34 / 3.36	2.31 / 3.30	2.26 / 3.19	2.21 / 3.12	2.15 / 3.00	2.11 / 2.92	2.07 / 2.84	2.02 / 2.76	2.00 / 2.70	1.96 / 2.63	1.94 / 2.60	1.88 / 2.49	19
20	4.35 / 8.10	3.49 / 5.85	3.10 / 4.94	2.87 / 4.43	2.71 / 4.10	2.60 / 3.87	2.52 / 3.71	2.45 / 3.56	2.40 / 3.45	2.35 / 3.37	2.31 / 3.30	2.28 / 3.23	2.23 / 3.13	2.18 / 3.05	2.12 / 2.94	2.08 / 2.86	2.04 / 2.77	1.99 / 2.69	1.96 / 2.63	1.92 / 2.56	1.90 / 2.53	1.84 / 2.42	20
21	4.32 / 8.02	3.47 / 5.78	3.07 / 4.87	2.84 / 4.37	2.68 / 4.04	2.57 / 3.81	2.49 / 3.65	2.42 / 3.51	2.37 / 3.40	2.32 / 3.31	2.28 / 3.24	2.25 / 3.17	2.20 / 3.07	2.15 / 2.99	2.09 / 2.88	2.05 / 2.80	2.00 / 2.72	1.96 / 2.63	1.93 / 2.58	1.89 / 2.51	1.87 / 2.47	1.81 / 2.36	21
22	4.30 / 7.94	3.44 / 5.72	3.05 / 4.82	2.82 / 4.31	2.66 / 3.99	2.55 / 3.76	2.47 / 3.59	2.40 / 3.45	2.35 / 3.35	2.30 / 3.26	2.26 / 3.18	2.23 / 3.12	2.18 / 3.02	2.13 / 2.94	2.07 / 2.83	2.03 / 2.75	1.98 / 2.67	1.93 / 2.58	1.91 / 2.53	1.87 / 2.46	1.84 / 2.42	1.78 / 2.31	22
23	4.28 / 7.88	3.42 / 5.66	3.03 / 4.76	2.80 / 4.26	2.64 / 3.94	2.53 / 3.71	2.45 / 3.54	2.38 / 3.41	2.32 / 3.30	2.28 / 3.21	2.24 / 3.14	2.20 / 3.07	2.14 / 2.97	2.10 / 2.89	2.04 / 2.78	2.00 / 2.70	1.96 / 2.62	1.91 / 2.53	1.88 / 2.48	1.84 / 2.41	1.82 / 2.37	1.76 / 2.26	23
24	4.26 / 7.82	3.40 / 5.61	3.01 / 4.72	2.78 / 4.22	2.62 / 3.90	2.51 / 3.67	2.43 / 3.50	2.36 / 3.36	2.30 / 3.25	2.26 / 3.17	2.22 / 3.09	2.18 / 3.03	2.13 / 2.93	2.09 / 2.85	2.02 / 2.74	1.98 / 2.66	1.94 / 2.58	1.89 / 2.49	1.86 / 2.44	1.82 / 2.36	1.80 / 2.33	1.73 / 2.21	24
25	4.24 / 7.77	3.38 / 5.57	2.99 / 4.68	2.76 / 4.18	2.60 / 3.86	2.49 / 3.63	2.41 / 3.46	2.34 / 3.32	2.28 / 3.21	2.24 / 3.13	2.20 / 3.05	2.16 / 2.99	2.11 / 2.89	2.06 / 2.81	2.00 / 2.70	1.96 / 2.62	1.92 / 2.54	1.87 / 2.45	1.84 / 2.40	1.80 / 2.32	1.77 / 2.29	1.71 / 2.17	25
26	4.22 / 7.72	3.37 / 5.53	2.98 / 4.64	2.74 / 4.14	2.59 / 3.82	2.47 / 3.59	2.39 / 3.42	2.32 / 3.29	2.27 / 3.17	2.22 / 3.09	2.18 / 3.02	2.15 / 2.96	2.10 / 2.86	2.05 / 2.77	1.99 / 2.66	1.95 / 2.58	1.90 / 2.50	1.85 / 2.41	1.82 / 2.36	1.78 / 2.28	1.76 / 2.25	1.69 / 2.13	26

TABLE A-6 (continued)

f_1 Degrees of Freedom (for greater mean square)

f_2	1	2	3	4	5	6	7	8	9	10	11	12	14	16	20	24	30	40	50	75	100	∞	f_2
27	4.21 / **7.68**	3.35 / **5.49**	2.96 / **4.60**	2.73 / **4.11**	2.57 / **3.79**	2.46 / **3.56**	2.37 / **3.39**	2.30 / **3.26**	2.25 / **3.14**	2.20 / **3.06**	2.16 / **2.98**	2.13 / **2.93**	2.08 / **2.83**	2.03 / **2.74**	1.97 / **2.63**	1.93 / **2.55**	1.88 / **2.47**	1.84 / **2.38**	1.80 / **2.33**	1.76 / **2.25**	1.74 / **2.21**	1.67 / **2.10**	27
28	4.20 / **7.64**	3.34 / **5.45**	2.95 / **4.57**	2.71 / **4.07**	2.56 / **3.76**	2.44 / **3.53**	2.36 / **3.36**	2.29 / **3.23**	2.24 / **3.11**	2.19 / **3.03**	2.15 / **2.95**	2.12 / **2.90**	2.06 / **2.80**	2.02 / **2.71**	1.96 / **2.60**	1.91 / **2.52**	1.87 / **2.44**	1.81 / **2.35**	1.78 / **2.30**	1.75 / **2.22**	1.72 / **2.18**	1.65 / **2.06**	28
29	4.18 / **7.60**	3.33 / **5.42**	2.93 / **4.54**	2.70 / **4.04**	2.54 / **3.73**	2.43 / **3.50**	2.35 / **3.33**	2.28 / **3.20**	2.22 / **3.08**	2.18 / **3.00**	2.14 / **2.92**	2.10 / **2.87**	2.05 / **2.77**	2.00 / **2.68**	1.94 / **2.57**	1.90 / **2.49**	1.85 / **2.41**	1.80 / **2.32**	1.77 / **2.27**	1.73 / **2.19**	1.71 / **2.15**	1.64 / **2.03**	29
30	4.17 / **7.56**	3.32 / **5.39**	2.92 / **4.51**	2.69 / **4.02**	2.53 / **3.70**	2.42 / **3.47**	2.34 / **3.30**	2.27 / **3.17**	2.21 / **3.06**	2.16 / **2.98**	2.12 / **2.90**	2.09 / **2.84**	2.04 / **2.74**	1.99 / **2.66**	1.93 / **2.55**	1.89 / **2.47**	1.84 / **2.38**	1.79 / **2.29**	1.76 / **2.24**	1.72 / **2.16**	1.69 / **2.13**	1.62 / **2.01**	30
32	4.15 / **7.50**	3.30 / **5.34**	2.90 / **4.46**	2.67 / **3.97**	2.51 / **3.66**	2.40 / **3.42**	2.32 / **3.25**	2.25 / **3.12**	2.19 / **3.01**	2.14 / **2.94**	2.10 / **2.86**	2.07 / **2.80**	2.02 / **2.70**	1.97 / **2.62**	1.91 / **2.51**	1.86 / **2.42**	1.82 / **2.34**	1.76 / **2.25**	1.74 / **2.20**	1.69 / **2.12**	1.67 / **2.08**	1.59 / **1.96**	32
34	4.13 / **7.44**	3.28 / **5.29**	2.88 / **4.42**	2.65 / **3.93**	2.49 / **3.61**	2.38 / **3.38**	2.30 / **3.21**	2.23 / **3.08**	2.17 / **2.97**	2.12 / **2.89**	2.08 / **2.82**	2.05 / **2.76**	2.00 / **2.66**	1.95 / **2.58**	1.89 / **2.47**	1.84 / **2.38**	1.80 / **2.30**	1.74 / **2.21**	1.71 / **2.15**	1.67 / **2.08**	1.64 / **2.04**	1.57 / **1.91**	34
36	4.11 / **7.39**	3.26 / **5.25**	2.86 / **4.38**	2.63 / **3.89**	2.48 / **3.58**	2.36 / **3.35**	2.28 / **3.18**	2.21 / **3.04**	2.15 / **2.94**	2.10 / **2.86**	2.06 / **2.78**	2.03 / **2.72**	1.98 / **2.62**	1.93 / **2.54**	1.87 / **2.43**	1.82 / **2.35**	1.78 / **2.26**	1.72 / **2.17**	1.69 / **2.12**	1.65 / **2.04**	1.62 / **2.00**	1.55 / **1.87**	36
38	4.10 / **7.35**	3.25 / **5.21**	2.85 / **4.34**	2.62 / **3.86**	2.46 / **3.54**	2.35 / **3.32**	2.26 / **3.15**	2.19 / **3.02**	2.14 / **2.91**	2.09 / **2.82**	2.05 / **2.75**	2.02 / **2.69**	1.96 / **2.59**	1.92 / **2.51**	1.85 / **2.40**	1.80 / **2.32**	1.76 / **2.22**	1.71 / **2.14**	1.67 / **2.08**	1.63 / **2.00**	1.60 / **1.97**	1.53 / **1.84**	38
40	4.08 / **7.31**	3.23 / **5.18**	2.84 / **4.31**	2.61 / **3.83**	2.45 / **3.51**	2.34 / **3.29**	2.25 / **3.12**	2.18 / **2.99**	2.12 / **2.88**	2.07 / **2.80**	2.04 / **2.73**	2.00 / **2.66**	1.95 / **2.56**	1.90 / **2.49**	1.84 / **2.37**	1.79 / **2.29**	1.74 / **2.20**	1.69 / **2.11**	1.66 / **2.05**	1.61 / **1.97**	1.59 / **1.94**	1.51 / **1.81**	40
42	4.07 / **7.27**	3.22 / **5.15**	2.83 / **4.29**	2.59 / **3.80**	2.44 / **3.49**	2.32 / **3.26**	2.24 / **3.10**	2.17 / **2.96**	2.11 / **2.86**	2.06 / **2.77**	2.02 / **2.70**	1.99 / **2.64**	1.94 / **2.54**	1.89 / **2.46**	1.82 / **2.35**	1.78 / **2.26**	1.73 / **2.17**	1.68 / **2.08**	1.64 / **2.02**	1.60 / **1.94**	1.57 / **1.91**	1.49 / **1.78**	42
44	4.06 / **7.24**	3.21 / **5.12**	2.82 / **4.26**	2.58 / **3.78**	2.43 / **3.46**	2.31 / **3.24**	2.23 / **3.07**	2.16 / **2.94**	2.10 / **2.84**	2.05 / **2.75**	2.01 / **2.68**	1.98 / **2.62**	1.92 / **2.52**	1.88 / **2.44**	1.81 / **2.32**	1.76 / **2.24**	1.72 / **2.15**	1.66 / **2.06**	1.63 / **2.00**	1.58 / **1.92**	1.56 / **1.88**	1.48 / **1.75**	44
46	4.05 / **7.21**	3.20 / **5.10**	2.81 / **4.24**	2.57 / **3.76**	2.42 / **3.44**	2.30 / **3.22**	2.22 / **3.05**	2.14 / **2.92**	2.09 / **2.82**	2.04 / **2.73**	2.00 / **2.66**	1.97 / **2.60**	1.91 / **2.50**	1.87 / **2.42**	1.80 / **2.30**	1.75 / **2.22**	1.71 / **2.13**	1.65 / **2.04**	1.62 / **1.98**	1.57 / **1.90**	1.54 / **1.86**	1.46 / **1.72**	46
48	4.04 / **7.19**	3.19 / **5.08**	2.80 / **4.22**	2.56 / **3.74**	2.41 / **3.42**	2.30 / **3.20**	2.21 / **3.04**	2.14 / **2.90**	2.08 / **2.80**	2.03 / **2.71**	1.99 / **2.64**	1.96 / **2.58**	1.90 / **2.48**	1.86 / **2.40**	1.79 / **2.28**	1.74 / **2.20**	1.70 / **2.11**	1.64 / **2.02**	1.61 / **1.96**	1.56 / **1.88**	1.53 / **1.84**	1.45 / **1.70**	48

TABLE A-6 (continued)

f_1 Degrees of Freedom (for greater mean square)

f_2	1	2	3	4	5	6	7	8	9	10	11	12	14	16	20	24	30	40	50	75	100	∞	f_2
50	4.03 / **7.17**	3.18 / **5.06**	2.79 / **4.20**	2.56 / **3.72**	2.40 / **3.41**	2.29 / **3.18**	2.20 / **3.02**	2.13 / **2.88**	2.07 / **2.78**	2.02 / **2.70**	1.98 / **2.62**	1.95 / **2.56**	1.90 / **2.46**	1.85 / **2.39**	1.78 / **2.26**	1.74 / **2.18**	1.69 / **2.10**	1.63 / **2.00**	1.60 / **1.94**	1.55 / **1.86**	1.52 / **1.82**	1.44 / **1.68**	50
55	4.02 / **7.12**	3.17 / **5.01**	2.78 / **4.16**	2.54 / **3.68**	2.38 / **3.37**	2.27 / **3.15**	2.18 / **2.98**	2.11 / **2.85**	2.05 / **2.75**	2.00 / **2.66**	1.97 / **2.59**	1.93 / **2.53**	1.88 / **2.43**	1.83 / **2.35**	1.76 / **2.23**	1.72 / **2.15**	1.67 / **2.06**	1.61 / **1.96**	1.58 / **1.90**	1.52 / **1.82**	1.50 / **1.78**	1.41 / **1.64**	55
60	4.00 / **7.08**	3.15 / **4.98**	2.76 / **4.13**	2.52 / **3.65**	2.37 / **3.34**	2.25 / **3.12**	2.17 / **2.95**	2.10 / **2.82**	2.04 / **2.72**	1.99 / **2.63**	1.95 / **2.56**	1.92 / **2.50**	1.86 / **2.40**	1.81 / **2.32**	1.75 / **2.20**	1.70 / **2.12**	1.65 / **2.03**	1.59 / **1.93**	1.56 / **1.87**	1.50 / **1.79**	1.48 / **1.74**	1.39 / **1.60**	60
65	3.99 / **7.04**	3.14 / **4.95**	2.75 / **4.10**	2.51 / **3.62**	2.36 / **3.31**	2.24 / **3.09**	2.15 / **2.93**	2.08 / **2.79**	2.02 / **2.70**	1.98 / **2.61**	1.94 / **2.54**	1.90 / **2.47**	1.85 / **2.37**	1.80 / **2.30**	1.73 / **2.18**	1.68 / **2.09**	1.63 / **2.00**	1.57 / **1.90**	1.54 / **1.84**	1.49 / **1.76**	1.46 / **1.71**	1.37 / **1.56**	65
70	3.98 / **7.01**	3.13 / **4.92**	2.74 / **4.08**	2.50 / **3.60**	2.35 / **3.29**	2.23 / **3.07**	2.14 / **2.91**	2.07 / **2.77**	2.01 / **2.67**	1.97 / **2.59**	1.93 / **2.51**	1.89 / **2.45**	1.84 / **2.35**	1.79 / **2.28**	1.72 / **2.15**	1.67 / **2.07**	1.62 / **1.98**	1.56 / **1.88**	1.53 / **1.82**	1.47 / **1.74**	1.45 / **1.69**	1.35 / **1.53**	70
80	3.96 / **6.96**	3.11 / **4.88**	2.72 / **4.04**	2.48 / **3.56**	2.33 / **3.25**	2.21 / **3.04**	2.12 / **2.87**	2.05 / **2.74**	1.99 / **2.64**	1.95 / **2.55**	1.91 / **2.48**	1.88 / **2.41**	1.82 / **2.32**	1.77 / **2.24**	1.70 / **2.11**	1.65 / **2.03**	1.60 / **1.94**	1.54 / **1.84**	1.51 / **1.78**	1.45 / **1.70**	1.42 / **1.65**	1.32 / **1.49**	80
100	3.94 / **6.90**	3.09 / **4.82**	2.70 / **3.98**	2.46 / **3.51**	2.30 / **3.20**	2.19 / **2.99**	2.10 / **2.82**	2.03 / **2.69**	1.97 / **2.59**	1.92 / **2.51**	1.88 / **2.43**	1.85 / **2.36**	1.79 / **2.26**	1.75 / **2.19**	1.68 / **2.06**	1.63 / **1.98**	1.57 / **1.89**	1.51 / **1.79**	1.48 / **1.73**	1.42 / **1.64**	1.39 / **1.59**	1.28 / **1.43**	100
∞	3.84 / **6.64**	2.99 / **4.60**	2.60 / **3.78**	2.37 / **3.32**	2.21 / **3.02**	2.09 / **2.80**	2.01 / **2.64**	1.94 / **2.51**	1.88 / **2.41**	1.83 / **2.32**	1.79 / **2.24**	1.75 / **2.18**	1.69 / **2.07**	1.64 / **1.99**	1.57 / **1.87**	1.52 / **1.79**	1.46 / **1.69**	1.40 / **1.59**	1.35 / **1.52**	1.28 / **1.41**	1.24 / **1.36**	1.00 / **1.00**	∞

Source: Reprinted by permission from *Statistical Methods* by George W. Snedecor and William G. Cochran, sixth edition. © 1967 by Iowa State University Press. Ames, Iowa.

TABLE A-7.   Square Roots (.001 to 1000)

To find the square root of a number N in this table, merely find N in the appropriate column headed "N." The square root of that number, $\sqrt{N}$, will appear immediately to the right in the column headed "$\sqrt{N}$." The use of this table is greatly facilitated by the inclusion of values between .001 and 1.00. For values that are not in the table, observe the rule of moving the decimal place in the answer to the right (left) one place for every two positions the decimal was moved right (left) in the original number. For example:

$$\sqrt{19} = 4.359 \qquad \text{(from table)}$$

$$\sqrt{190} = 13.784 \qquad \text{(from table)}$$

$$\sqrt{1900} = 43.59$$

$$\sqrt{19000} = 137.84$$

$$\sqrt{190000} = 435.9$$

$$\sqrt{1900000} = 1378.4$$

TABLE A-7 (continued)

N	$\sqrt{N}$	N	$\sqrt{N}$	N	$\sqrt{N}$	N	$\sqrt{N}$
.001	.0316	.051	.2258	.101	.3178	.151	.3886
.002	.0447	.052	.2280	.102	.3194	.152	.3899
.003	.0548	.053	.2302	.103	.3209	.153	.3912
.004	.0632	.054	.2324	.104	.3225	.154	.3924
.005	.0707	.055	.2345	.105	.3240	.155	.3937
.006	.0775	.056	.2366	.106	.3256	.156	.3950
.007	.0837	.057	.2387	.107	.3271	.157	.3962
.008	.0894	.058	.2408	.108	.3286	.158	.3975
.009	.0949	.059	.2429	.109	.3302	.159	.3987
.010	.1000	.060	.2449	.110	.3317	.160	.4000
.011	.1049	.061	.2470	.111	.3332	.161	.4012
.012	.1095	.062	.2490	.112	.3347	.162	.4025
.013	.1140	.063	.2510	.113	.3362	.163	.4037
.014	.1183	.064	.2530	.114	.3376	.164	.4050
.015	.1225	.065	.2550	.115	.3391	.165	.4062
.016	.1265	.066	.2569	.116	.3406	.166	.4074
.017	.1304	.067	.2588	.117	.3421	.167	.4087
.018	.1342	.068	.2608	.118	.3435	.168	.4099
.019	.1378	.069	.2627	.119	.3450	.169	.4111
.020	.1414	.070	.2646	.120	.3464	.170	.4123
.021	.1449	.071	.2665	.121	.3479	.171	.4135
.022	.1483	.072	.2683	.122	.3493	.172	.4147
.023	.1517	.073	.2702	.123	.3507	.173	.4159
.024	.1549	.074	.2720	.124	.3521	.174	.4171
.025	.1581	.075	.2739	.125	.3536	.175	.4183
.026	.1612	.076	.2757	.126	.3550	.176	.4195
.027	.1643	.077	2775	.127	.3564	.177	.4207
.028	.1673	.078	.2793	.128	.3578	.178	.4219
.029	.1703	.079	.2811	.129	.3592	.179	.4231
.030	.1732	.080	.2828	.130	.3606	.180	.4243
.031	.1761	.081	.2846	.131	.3619	.181	.4254
.032	.1789	.082	.2864	.132	.3633	.182	.4266
.033	.1817	.083	.2881	.133	.3647	.183	.4278
.034	.1844	.084	.2898	.134	.3661	.184	.4290
.035	.1871	.085	.2915	.135	.3674	.185	.4301
.036	.1897	.086	.2933	.136	.3688	.186	.4313
.037	.1924	.087	.2950	.137	.3701	.187	.4324
.038	.1949	.088	.2966	.138	.3715	.188	.4336
.039	.1975	.089	.2983	.139	.3728	.189	.4347
.040	.2000	.090	.3000	.140	.3742	.190	.4359
.041	.2025	.091	.3017	.141	.3755	.191	.4370
.042	.2049	.092	.3033	.142	.3768	.192	.4382
.043	.2074	.093	.3050	.143	.3782	.193	.4393
.044	.2098	.094	.3066	.144	.3795	.194	.4405
.045	.2121	.095	.3082	.145	.3808	.195	.4416
.046	.2145	.096	.3098	.146	.3821	.196	.4427
.047	.2168	.097	.3114	.147	.3834	.197	.4438
.048	.2191	.098	.3130	.148	.3847	.198	.4450
.049	.2214	.099	.3146	.149	.3860	.199	.4461
.050	.2236	.100	.3162	.150	.3873	.200	.4472

TABLE A-7 (continued)

N	$\sqrt{N}$	N	$\sqrt{N}$	N	$\sqrt{N}$	N	$\sqrt{N}$
.201	.4483	.251	.5010	.301	.5486	.351	.5925
.202	.4494	.252	.5020	.302	.5495	.352	.5933
.203	.4506	.253	.5030	.303	.5505	.353	.5941
.204	.4517	.254	.5040	.304	.5514	.354	.5950
.205	.4528	.255	.5050	.305	.5523	.355	.5958
.206	.4539	.256	.5060	.306	.5532	.356	.5967
.207	.4550	.257	.5070	.307	.5541	.357	.5975
.208	.4561	.258	.5079	.308	.5550	.358	.5983
.209	.4572	.259	.5089	.309	.5559	.359	.5992
.210	.4583	.260	.5099	.310	.5568	.360	.6000
.211	.4593	.261	.5109	.311	.5577	.361	.6008
.212	.4604	.262	.5119	.312	.5586	.362	.6017
.213	.4615	.263	.5128	.313	.5595	.363	.6025
.214	.4626	.264	.5138	.314	.5604	.364	.6033
.215	.4637	.265	.5148	.315	.5612	.365	.6042
.216	.4648	.266	.5158	.316	.5621	.366	.6050
.217	.4658	.267	.5167	.317	.5630	.367	.6058
.218	.4669	.268	.5177	.318	.5639	.368	.6066
.219	.4680	.269	.5187	.319	.5648	.369	.6075
.220	.4690	.270	.5196	.320	.5657	.370	.6083
.221	.4701	.271	.5206	.321	.5666	.371	.6091
.222	.4712	.272	.5215	.322	.5675	.372	.6099
.223	.4722	.273	.5225	.323	.5683	.373	.6107
.224	.4733	.274	.5235	.324	.5692	.374	.6116
.225	.4743	.275	.5244	.325	.5701	.375	.6124
.226	.4754	.276	.5254	.326	.5710	.376	.6132
.227	.4764	.277	.5263	.327	.5718	.377	.6140
.228	.4775	.278	.5273	.328	.5727	.378	.6148
.229	.4785	.279	.5282	.329	.5736	.379	.6156
.230	.4796	.280	.5292	.330	.5745	.380	.6164
.231	.4806	.281	.5301	.331	.5753	.381	.6173
.232	.4817	.282	.5310	.332	.5762	.382	.6181
.233	.4827	.283	.5320	.333	.5771	.383	.6189
.234	.4837	.284	.5329	.334	.5779	.384	.6197
.235	.4848	.285	.5339	.335	.5788	.385	.6205
.236	.4858	.286	.5348	.336	.5797	.386	.6213
.237	.4868	.287	.5357	.337	.5805	.387	.6221
.238	.4879	.288	.5367	.338	.5814	.388	.6229
.239	.4889	.289	.5376	.339	.5822	.389	.6237
.240	.4899	.290	.5385	.340	.5831	.390	.6245
.241	.4909	.291	.5394	.341	.5840	.391	.6253
.242	.4919	.292	.5404	.342	.5848	.392	.6261
.243	.4930	.293	.5413	.343	.5857	.393	.6269
244	.4940	.294	.5422	.344	.5865	.394	.6277
.245	.4950	.295	.5431	.345	.5874	.395	.6285
.246	.4960	.296	.5441	.346	.5882	.396	.6293
.247	.4970	.297	.5450	.347	.5891	.397	.6301
.248	.4980	.298	.5459	.348	.5899	.398	.6309
.249	.4990	.299	.5468	.349	.5908	.399	.6317
.250	.5000	.300	.5477	.350	.5916	.400	.6325

TABLE A-7 (continued)

N	$\sqrt{N}$	N	$\sqrt{N}$	N	$\sqrt{N}$	N	$\sqrt{N}$
.401	.6332	.451	.6716	.501	.7078	.551	.7423
.402	.6340	.452	.6723	.502	.7085	.552	.7430
.403	.6348	.453	.6731	.503	.7092	.553	.7436
.404	.6356	.454	.6738	.504	.7099	.554	.7443
.405	.6364	.455	.6745	.505	.7106	.555	.7450
.406	.6372	.456	.6753	.506	.7113	.556	.7457
.407	.6380	.457	.6760	.507	.7120	.557	.7463
.408	.6387	.458	.6768	.508	.7127	.558	.7470
.409	.6395	.459	.6775	.509	.7134	.559	.7477
.410	.6403	.460	.6782	.510	.7141	.560	.7483
.411	.6411	.461	.6790	.511	.7148	.561	.7490
.412	.6419	.462	.6797	.512	.7155	.562	.7497
.413	.6427	.463	.6804	.513	.7162	.563	.7503
.414	.6434	.464	.6812	.514	.7169	.564	.7510
.415	.6442	.465	.6819	.515	.7176	.565	.7517
.416	.6450	.466	.6826	.516	.7183	.566	.7523
.417	.6458	.467	.6834	.517	.7190	.567	.7530
.418	.6465	.468	.6841	.518	.7197	.568	.7537
.419	.6473	.469	.6848	.519	.7204	.569	.7543
.420	.6481	.470	.6856	.520	.7211	.570	.7550
.421	.6488	.471	.6863	.521	.7218	.571	.7556
.422	.6496	.472	.6870	.522	.7225	.572	.7563
.423	.6504	.473	.6877	.523	.7232	.573	.7570
.424	.6512	.474	.6885	.524	.7239	.574	.7576
.425	.6519	.475	.6892	.525	.7246	.575	.7583
.426	.6527	.476	.6899	.526	.7253	.576	.7589
.427	.6535	.477	.6907	.527	.7259	.577	.7596
.428	.6542	.478	.6914	.528	.7266	.578	.7603
.429	.6550	.479	.6921	.529	.7273	.579	.7609
.430	.6557	.480	.6928	.530	.7280	.580	.7616
.431	.6565	.481	.6935	.531	.7287	.581	.7622
.432	.6573	.482	.6943	.532	.7294	.582	.7629
.433	.6580	.483	.6950	.533	.7301	.583	.7635
.434	.6588	.484	.6957	.534	.7308	.584	.7642
.435	6595	.485	.6964	.535	.7314	.585	.7649
.436	.6603	.486	.6971	.536	.7321	.586	.7655
.437	.6611	.487	.6979	.537	.7328	.587	.7662
.438	.6618	.488	.6986	.538	.7335	.588	.7668
.439	.6626	.489	.6993	.539	.7342	.589	.7675
.440	.6633	.490	.7000	.540	.7348	.590	.7681
.441	.6641	.491	.7007	.541	.7355	.591	.7688
.442	.6648	.492	.7014	.542	.7362	.592	.7694
.443	.6656	.493	.7021	.543	.7369	.593	.7701
.444	.6663	.494	.7029	.544	.7376	.594	.7707
.445	.6671	.495	.7036	.545	.7382	.595	.7714
.446	.6678	.496	.7043	.546	.7389	.596	.7720
.447	.6686	.497	.7050	.547	.7396	.597	.7727
.448	.6693	.498	.7057	.548	.7403	.598	.7733
.449	.6701	.499	.7064	.549	.7409	.599	.7740
.450	.6708	.500	.7071	.550	.7416	.600	.7746

TABLE A-7 (continued)

N	$\sqrt{N}$	N	$\sqrt{N}$	N	$\sqrt{N}$	N	$\sqrt{N}$
.601	.7752	.651	.8068	.701	.8373	.751	.8666
.602	.7759	.652	.8075	.702	.8379	.752	.8672
.603	.7765	.653	.8081	.703	.8385	.753	.8678
.604	.7772	.654	.8087	.704	.8390	.754	.8683
.605	.7778	.655	.8093	.705	.8396	.755	.8689
.606	.7785	.656	.8099	.706	.8402	.756	.8695
.607	.7791	.657	.8106	.707	.8408	.757	.8701
.608	.7797	.658	.8112	.708	.8414	.758	.8706
.609	.7804	.659	.8118	.709	.8420	.759	.8712
.610	.7810	.660	.8124	.710	.8426	.760	.8718
.611	.7817	.661	.8130	.711	.8432	.761	.8724
.612	.7823	.662	.8136	.712	.8438	.762	.8729
.613	.7829	.663	.8142	.713	.8444	.763	.8735
.614	.7836	.664	.8149	.714	.8450	.764	.8741
.615	.7842	.665	.8155	.715	.8456	.765	.8746
.616	.7849	.666	.8161	.716	.8462	.766	.8752
.617	.7855	.667	.8167	.717	.8468	.767	.8758
.618	.7861	.668	.8173	.718	.8473	.768	.8764
.619	.7868	.669	.8179	.719	.8479	.769	.8769
.620	.7874	.670	.8185	.720	.8485	.770	.8775
.621	.7880	.671	.8191	.721	.8491	.771	.8781
.622	.7887	.672	.8198	.722	.8497	.772	.8786
.623	.7893	.673	.8204	.723	.8503	.773	.8792
.624	.7899	.674	.8210	.724	.8509	.774	.8798
.625	.7906	.675	.8216	.725	.8515	.775	.8803
.626	.7912	.676	.8222	.726	.8521	.776	.8809
.627	.7918	.677	.8228	.727	.8526	.777	.8815
.628	.7925	.678	.8234	.728	.8532	.778	.8820
.629	.7931	.679	.8240	.729	.8538	.779	.8826
.630	.7937	.680	.8246	.730	.8544	.780	.8832
.631	.7944	.681	.8252	.731	.8550	.781	.8837
.632	.7950	.682	.8258	.732	.8556	.782	.8843
.633	.7956	.683	.8264	.733	.8562	.783	.8849
.634	.7962	.684	.8270	.734	.8567	.784	.8854
.635	.7969	.685	.8276	.735	.8573	.785	.8860
.636	.7975	.686	.8283	.736	.8579	.786	.8866
.637	.7981	.687	.8289	.737	.8585	.787	.8871
.638	.7987	.688	.8295	.738	.8591	.788	.8877
.639	.7994	.689	.8301	.739	.8597	.789	.8883
.640	.8000	.690	.8307	.740	.8602	.790	.8888
.641	.8006	.691	.8313	.741	.8608	.791	.8894
.542	.8012	.692	.8319	.742	.8614	.792	.8899
.643	.8019	.693	.8325	.743	.8620	.793	.8905
.644	.8025	.694	.8331	.744	.8626	.794	.8911
.645	.8031	.695	.8337	.745	.8631	.795	.8916
.646	.8037	.696	.8343	.746	.8637	.796	.8922
.647	.8044	.697	.8349	.747	.8643	.797	.8927
.648	.8050	.698	.8355	.748	.8649	.798	.8933
.649	.8056	.699	.8361	.749	.8654	.799	.8939
.650	.8062	.700	.8367	.750	.8660	.800	.8944

TABLE A-7 (continued)

N	$\sqrt{N}$	N	$\sqrt{N}$	N	$\sqrt{N}$	N	$\sqrt{N}$
.801	.8950	.851	.9225	.901	.9492	.951	.9752
.802	.8955	.852	.9230	.902	.9497	.952	.9757
.803	.8961	.853	.9236	.903	.9503	.953	.9762
.804	.8967	.854	.9241	.904	.9508	.954	.9767
.805	.8972	.855	.9247	.905	.9513	.955	.9772
.806	.8978	.856	.9252	.906	.9518	.956	.9778
.807	.8983	.857	.9257	.907	.9524	.957	.9783
.808	.8989	.858	.9263	.908	.9529	.958	.9788
.809	.8994	.859	.9268	.909	.9534	.959	.9793
.810	.9000	.860	.9274	.910	.9539	.960	.9798
.811	.9006	.861	.9279	.911	.9545	.961	.9803
.812	.9011	.862	.9284	.912	.9550	.962	.9808
.813	.9017	.863	.9290	.913	.9555	.963	.9813
.814	.9022	.864	.9295	.914	.9560	.964	.9818
.815	.9028	.865	.9301	.915	.9566	.965	.9823
.816	.9033	.866	.9306	.916	.9571	.966	.9829
.817	.9039	.867	.9311	.917	.9576	.967	.9834
.818	.9044	.868	.9317	.918	.9581	.968	.9839
.819	.9050	.869	.9322	.919	.9586	.969	.9844
.820	.9055	.870	.9327	.920	.9592	.970	.9849
.821	.9061	.871	.9333	.921	.9597	.971	.9854
.822	.9066	.872	.9338	.922	.9602	.972	.9859
.823	.9072	.873	.9343	.923	.9607	.973	.9864
.824	.9077	.874	.9349	.924	.9612	.974	.9869
.825	.9083	.875	.9354	.925	.9618	.975	.9874
.826	.9088	.876	.9359	.926	.9623	.976	.9879
.827	.9094	.877	.9365	.927	.9628	.977	.9884
.828	.9099	.878	.9370	.928	.9633	.978	.9889
.829	.9105	.879	.9375	.929	.9638	.979	.9894
.830	.9110	.880	.9381	.930	.9644	.980	.9899
.831	.9116	.881	.9386	.931	.9649	.981	.9905
.832	.9121	.882	.9391	.932	.9654	.982	.9910
.833	.9127	.883	.9397	.933	.9659	.983	.9915
.834	.9132	.884	.9402	.934	.9664	.984	.9920
.835	.9138	.885	.9407	.935	.9670	.985	.9925
.836	.9143	.886	.9413	.936	.9675	.986	.9930
.837	.9149	.887	.9418	.937	.9680	.987	.9935
.838	.9154	.888	.9423	.938	.9685	.988	.9940
.839	.9160	.889	.9429	.939	.9690	.989	.9945
.840	.9165	.890	.9434	.940	.9695	.990	.9950
.841	.9171	.891	.9439	.941	.9701	.991	.9955
.842	.9176	.892	.9445	.942	.9706	.992	.9960
.843	.9182	.893	.9450	.943	.9711	.993	.9965
.844	.9187	.894	.9455	.944	.9716	.994	.9970
.845	.9192	.895	.9460	.945	.9721	.995	.9975
.846	.9198	.896	.9466	.946	.9726	.996	.9980
.847	.9203	.897	.9471	.947	.9731	.997	.9985
.848	.9209	.898	.9476	.948	.9737	.998	.9990
.849	.9214	.899	.9482	.949	.9742	.999	.9995
.850	.9220	.900	.9487	.950	.9747	1.000	1.0000

TABLE A-7 (continued)

N	$\sqrt{N}$	N	$\sqrt{N}$	N	$\sqrt{N}$	N	$\sqrt{N}$
1	1.000	51	7.141	101	10.050	151	12.288
2	1.414	52	7.211	102	10.100	152	12.329
3	1.732	53	7.280	103	10.149	153	12.369
4	2.000	54	7.348	104	10.198	154	12.410
5	2.236	55	7.416	105	10.247	155	12.450
6	2.449	56	7.483	106	10.296	156	12.490
7	2.646	57	7.550	107	10.344	157	12.530
8	2.828	58	7.616	108	10.392	158	12.570
9	3.000	59	7.681	109	10.440	159	12.610
10	3.162	60	7.746	110	10.488	160	12.649
11	3.317	61	7.810	111	10.536	161	12.689
12	3.464	62	7.874	112	10.583	162	12.728
13	3.606	63	7.937	113	10.630	163	12.767
14	3.742	64	8.000	114	10.677	164	12.806
15	3.873	65	8.062	115	10.724	165	12.845
16	4.000	66	8.124	116	10.770	166	12.884
17	4.123	67	8.185	117	10.817	167	12.923
18	4.243	68	8.246	118	10.863	168	12.961
19	4.359	69	8.307	119	10.909	169	13.000
20	4.472	70	8.367	120	10.954	170	13.038
21	4.583	71	8.426	121	11.000	171	13.077
22	4.690	72	8.485	122	11.045	172	13.115
23	4.796	73	8.544	123	11.091	173	13.153
24	4.899	74	8.602	124	11.136	174	13.191
25	5.000	75	8.660	125	11.180	175	13.229
26	5.099	76	8.718	126	11.225	176	13.266
27	5.196	77	8.775	127	11.269	177	13.304
28	5.292	78	8.832	128	11.314	178	13.342
29	5.385	79	8.888	129	11.358	179	13.379
30	5.477	80	8.944	130	11.402	180	13.416
31	5.568	81	9.000	131	11.446	181	13.454
32	5.657	82	9.055	132	11.489	182	13.491
33	5.745	83	9.110	133	11.533	183	13.528
34	5.831	84	9.165	134	11.576	184	13.565
35	5.916	85	9.220	135	11.619	185	13.601
36	6.000	86	9.274	136	11.662	186	13.638
37	6.083	87	9.327	137	11.705	187	13.675
38	6.164	88	9.381	138	11.747	188	13.711
39	6.245	89	9.434	139	11.790	189	13.748
40	6.325	90	9.487	140	11.832	190	13.784
41	6.403	91	9.539	141	11.874	191	13.820
42	6.481	92	9.592	142	11.916	192	13.856
43	6.557	93	9.644	143	11.958	193	13.892
44	6.633	94	9.695	144	12.000	194	13.928
45	6.708	95	9.747	145	12.042	195	13.964
46	6.782	96	9.798	146	12.083	196	14.000
47	6.856	97	9.849	147	12.124	197	14.036
48	6.928	98	9.899	148	12.166	198	14.071
49	7.000	99	9.950	149	12.207	199	14.107
50	7.071	100	10.000	150	12.247	200	14.142

TABLE A-7 (continued)

N	$\sqrt{N}$	N	$\sqrt{N}$	N	$\sqrt{N}$	N	$\sqrt{N}$
201	14.177	251	15.843	301	17.349	351	18.735
202	14.213	252	15.875	302	17.378	352	18.762
203	14.248	253	15.906	303	17.407	353	18.788
204	14.283	254	15.937	304	17.436	354	18.815
205	14.318	255	15.969	305	17.464	355	18.841
206	14.353	256	16.000	306	17.493	356	18.868
207	14.387	257	16.031	307	17.521	357	18.894
208	14.422	258	16.062	308	17.550	358	18.921
209	14.457	259	16.093	309	17.578	359	18.947
210	14.491	260	16.125	310	17.607	360	18.974
211	14.526	261	16.155	311	17.635	361	19.000
212	14.560	262	16.186	312	17.664	362	19.026
213	14.595	263	16.217	313	17.692	363	19.053
214	14.629	264	16.248	314	17.720	364	19.079
215	14.663	265	16.279	315	17.748	365	19.105
216	14.697	266	16.310	316	17.776	366	19.131
217	14.731	267	16.340	317	17.804	367	19.157
218	14.765	268	16.371	318	17.833	368	19.183
219	14.799	269	16.401	319	17.861	369	19.209
220	14.832	270	16.432	320	17.889	370	19.235
221	14.866	271	16.462	321	17.916	371	19.261
222	14.900	272	16.492	322	17.944	372	19.287
223	14.933	273	16.523	323	17.972	373	19.313
224	14.967	274	16.553	324	18.000	374	19.339
225	15.000	275	16.583	325	18.028	375	19.365
226	15.033	276	16.613	326	18.055	376	19.391
227	15.067	277	16.643	327	18.083	377	19.416
228	15.100	278	16.673	328	18.111	378	19.442
229	15.133	279	16.703	329	18.138	379	19.468
230	15.166	280	16.733	330	18.166	380	19.494
231	15.199	281	16.763	331	18.193	381	19.519
232	15.232	282	16.793	332	18.221	382	19.545
233	15.264	283	16.823	333	18.248	383	19.570
234	15.297	284	16.852	334	18.276	384	19.596
235	15.330	285	16.882	335	18.303	385	19.621
236	15.362	286	16.912	336	18.330	386	19.647
237	15.395	287	16.941	337	18.358	387	19.672
238	15.427	288	16.971	338	18.385	388	19.698
239	15.460	289	17.000	339	18.412	389	19.723
240	15.492	290	17.029	340	18.439	390	19.748
241	15.524	291	17.059	341	18.466	391	19.774
242	15.556	292	17.088	342	18.493	392	19.799
243	15.588	293	17.117	343	18.520	393	19.824
244	15.620	294	17.146	344	18.547	394	19.849
245	15.652	295	17.176	345	18.574	395	19.875
246	15.684	296	17.205	346	18.601	396	19.900
247	15.716	297	17.234	347	18.628	397	19.925
248	15.748	298	17.263	348	18.655	398	19.950
249	15.780	299	17.292	349	18.682	399	19.975
250	15.811	300	17.321	350	18.708	400	20.000

TABLE A-7 (continued)

N	$\sqrt{N}$	N	$\sqrt{N}$	N	$\sqrt{N}$	N	$\sqrt{N}$
401	20.025	451	21.237	501	22.383	551	23.473
402	20.050	452	21.260	502	22.405	552	23.495
403	20.075	453	21.284	503	22.428	553	23.516
404	20.100	454	21.307	504	22.450	554	23.537
405	20.125	455	21.331	505	22.472	555	23.558
406	20.149	456	21.354	506	22.494	556	23.580
407	20.174	457	21.378	507	22.517	557	23.601
408	20.199	458	21.401	508	22.539	558	23.622
409	20.224	459	21.424	509	22.561	559	23.643
410	20.248	460	21.448	510	22.583	560	23.664
411	20.273	461	21.471	511	22.605	561	23.685
412	20.298	462	21.494	512	22.627	562	23.707
413	20.322	463	21.517	513	22.650	563	23.728
414	20.347	464	21.541	514	22.672	564	23.749
415	20.372	465	21.564	515	22.694	565	23.770
416	20.396	466	21.587	516	22.716	566	23.791
417	20.421	467	21.610	517	22.738	567	23.812
418	20.445	468	21.633	518	22.760	568	23.833
419	20.469	469	21.656	519	22.782	569	23.854
420	20.494	470	21.679	520	22.804	570	23.875
421	20.518	471	21.703	521	22.825	571	23.896
422	20.543	472	21.726	522	22.847	572	23.917
423	20.567	473	21.749	523	22.869	573	23.937
424	20.591	474	21.772	524	22.891	574	23.958
425	20.616	475	21.794	525	22.913	575	23.979
426	20.640	476	21.817	526	22.935	576	24.000
427	20.664	477	21.840	527	22.956	577	24.021
428	20.688	478	21.863	528	22.978	578	24.042
429	20.712	479	21.886	529	23.000	579	24.062
430	20.736	480	21.909	530	23.022	580	24.083
431	20.761	481	21.932	531	23.043	581	24.104
432	20.785	482	21.954	532	23.065	582	24.125
433	20.809	483	21.977	533	23.087	583	24.145
434	20.833	484	22.000	534	23.108	584	24.166
435	20.857	485	22.023	535	23.130	585	24.187
436	20.881	486	22.045	536	23.152	586	24.207
437	20.905	487	22.068	537	23.173	587	24.228
438	20.928	488	22.091	538	23.195	588	24.249
439	20.952	489	22.113	539	23.216	589	24.269
440	20.976	490	22.136	540	23.238	590	24.290
441	21.000	491	22.159	541	23.259	591	24.310
442	21.024	492	22.181	542	23.281	592	24.331
443	21.048	493	22.204	543	23.302	593	24.352
444	21.071	494	22.226	544	23.324	594	24.372
445	21.095	495	22.249	545	23.345	595	24.393
446	21.119	496	22.271	546	23.367	596	24.413
447	21.142	497	22.293	547	23.388	597	24.434
448	21.166	498	22.316	548	23.409	598	24.454
449	21.190	499	22.338	549	23.431	599	24.474
450	21.213	500	22.361	550	23.452	600	24.495

TABLE A-7 (continued)

N	$\sqrt{N}$	N	$\sqrt{N}$	N	$\sqrt{N}$	N	$\sqrt{N}$
601	24.515	651	25.515	701	26.476	751	27.404
602	24.536	652	25.534	702	26.495	752	27.423
603	24.556	653	25.554	703	26.514	753	27.441
604	24.576	654	25.573	704	26.533	754	27.459
605	24.597	655	25.593	705	26.552	755	27.477
606	24.617	656	25.612	706	26.571	756	27.495
607	24.637	657	25.632	707	26.589	757	27.514
608	24.658	658	25.652	708	26.608	758	27.532
609	24.678	659	25.671	709	26.627	759	27.550
610	24.698	660	25.690	710	26.646	760	27.568
611	24.718	661	25.710	711	26.665	761	27.586
612	24.739	662	25.729	712	26.683	762	27.604
613	24.759	663	25.749	713	26.702	763	27.622
614	24.779	664	25.768	714	26.721	764	27.641
615	24.799	665	25.788	715	26.739	765	27.659
616	24.819	666	25.807	716	26.758	766	27.677
617	24.839	667	25.826	717	26.777	767	27.695
618	24.860	668	25.846	718	26.796	768	27.713
619	24.880	669	25.865	719	26.814	769	27.731
620	24.900	670	25.884	720	26.833	770	27.749
621	24.920	671	25.904	721	26.851	771	27.767
622	24.940	672	25.923	722	26.870	772	27.785
623	24.960	673	25.942	723	26.889	773	27.803
624	24.980	674	25.962	724	26.907	774	27.821
625	25.000	675	25.981	725	26.926	775	27.839
626	25.020	676	26.000	726	26.944	776	27.857
627	25.040	677	26.019	727	26.963	777	27.875
628	25.060	678	26.038	728	26.981	778	27.893
629	25.080	679	26.058	729	27.000	779	27.911
630	25.100	680	26.077	730	27.019	780	27.928
631	25.120	681	26.096	731	27.037	781	27.946
632	25.140	682	26.115	732	27.055	782	27.964
633	25.159	683	26.134	733	27.074	783	27.982
634	25.179	684	26.153	734	27.092	784	28.000
635	25.199	685	26.173	735	27.111	785	28.018
636	25.219	686	26.192	736	27.129	786	28.036
637	25.239	687	26.211	737	27.148	787	28.054
638	25.259	688	26.230	738	27.166	788	28.071
639	25.278	689	26.249	739	27.185	789	28.089
640	25.298	690	26.268	740	27.203	790	28.107
641	25.318	691	26.287	741	27.221	791	28.125
642	25.338	692	26.306	742	27.240	792	28.142
643	25.357	693	26.325	743	27.258	793	28.160
644	25.377	694	26.344	744	27.276	794	28.178
645	25.397	695	26.363	745	27.295	795	28.196
646	25.417	696	26.382	746	27.313	796	28.213
647	25.436	697	26.401	747	27.331	797	28.231
648	25.456	698	26.420	748	27.350	798	28.249
649	25.475	699	26.439	749	27.368	799	28.267
650	25.495	700	26.458	750	27.386	800	28.284

TABLE A-7 (continued)

N	$\sqrt{N}$	N	$\sqrt{N}$	N	$\sqrt{N}$	N	$\sqrt{N}$
801	28.302	851	29.172	901	30.017	951	30.838
802	28.320	852	29.189	902	30.033	952	30.854
803	28.337	853	29.206	903	30.050	953	30.871
804	28.355	854	29.223	904	30.067	954	30.887
805	28.373	855	29.240	905	30.083	955	30.903
806	28.390	856	29.257	906	30.100	956	30.919
807	28.408	857	29.275	907	30.116	957	30.935
808	28.425	858	29.292	908	30.133	958	30.952
809	28.443	859	29.309	909	30.150	959	30.968
810	28.460	860	29.326	910	30.166	960	30.984
811	28.478	861	29.343	911	30.183	961	31.000
812	28.496	862	29.360	912	30.199	962	31.016
813	28.513	863	29.377	913	30.216	963	31.032
814	28.531	864	29.394	914	30.232	964	31.048
815	28.548	865	29.411	915	30.249	965	31.064
816	28.566	866	29.428	916	30.265	966	31.081
817	28.583	867	29.445	917	30.282	967	31.097
818	28.601	868	29.462	918	30.299	968	31.113
819	28.618	869	29.479	919	30.315	969	31.129
820	28.636	870	29.496	920	30.332	970	31.145
821	28.653	871	29.513	921	30.348	971	31.161
822	28.671	872	29.530	922	30.364	972	31.177
823	28.688	873	29.547	923	30.381	973	31.193
824	28.705	874	29.563	924	30.397	974	31.209
825	28.723	875	29.580	925	30.414	975	31.225
826	28.740	876	29.597	926	30.430	976	31.241
827	28.758	877	29.614	927	30.447	977	31.257
828	28.775	878	29.631	928	30.463	978	31.273
829	28.792	879	29.648	929	30.480	979	31.289
830	28.810	880	29.665	930	30.496	980	31.305
831	28.827	881	29.682	931	30.512	981	31.321
832	28.844	882	29.698	932	30.529	982	31.337
833	28.862	883	29.715	933	30.545	983	31.353
834	28.879	884	29.732	934	30.561	984	31.369
835	28.896	885	29.749	935	30.578	985	31.385
836	28.914	886	29.766	936	30.594	986	31.401
837	28.931	887	29.783	937	30.610	987	31.417
838	28.948	888	29.799	938	30.627	988	31.432
839	28.965	889	29.816	939	30.643	989	31.448
840	28.983	890	29.833	940	30.659	990	31.464
841	29.000	891	29.850	941	30.676	991	31.480
842	29.017	892	29.866	942	30.692	992	31.496
843	29.034	893	29.883	943	30.708	993	31.512
844	29.052	894	29.900	944	30.725	994	31.528
845	29.069	895	29.917	945	30.741	995	31.544
846	29.086	896	29.933	946	30.757	996	31.559
847	29.103	897	29.950	947	30.773	997	31.575
848	29.120	898	29.967	948	30.790	998	31.591
849	29.138	899	29.983	949	30.806	999	31.607
850	29.155	900	30.000	950	30.822	1000	31.623

INDEX